Lecture Notes in Computer Science 9396

Commenced Publication in 1973
Founding and Former Series Editors:
Gerhard Goos, Juris Hartmanis, and Jan van Leeuwen

More information about this series at http://www.springer.com/series/7409

Florian Daniel · Oscar Diaz (Eds.)

Current Trends in Web Engineering

15th International Conference, ICWE 2015 Workshops
NLPIT, PEWET, SoWEMine
Rotterdam, The Netherlands, June 23–26, 2015
Revised Selected Papers

 Springer

Editors
Florian Daniel
Università di Trento
Povo, Trento
Italy

Oscar Diaz
Universidad del Pais Vasco
San Sebastian
Spain

ISSN 0302-9743   ISSN 1611-3349   (electronic)
Lecture Notes in Computer Science
ISBN 978-3-319-24799-1   ISBN 978-3-319-24800-4   (eBook)
DOI 10.1007/978-3-319-24800-4

Library of Congress Control Number: 2015950045

LNCS Sublibrary: SL3 – Information Systems and Applications, incl. Internet/Web, and HCI

Springer Cham Heidelberg New York Dordrecht London

Printed on acid-free paper

Springer International Publishing AG Switzerland is part of Springer Science+Business Media
(www.springer.com)

Foreword

Workshops strive to be places for open speech and respectful dissension, where preliminary ideas can be discussed and opposite views peacefully compared. If this is the aim of workshops, no place but the hometown of Erasmus of Rotterdam signifies this spirit. This leading humanist stands for the critical and open mind that should characterize workshop sessions. While critical about the abuses within the Catholic Church, he kept a distance from Martin Luther's reformist ideas, emphasizing a middle way with a deep respect for traditional faith, piety, and grace, rejecting Luther's emphasis on faith alone. Though far from the turbulent days of the XV century, Web Engineering is a battlefield where the irruption of new technologies challenges not only software architectures but also established social and business models. This makes workshops not mere co-located events of a conference but an *essential* part of it, allowing one to feel the pulse of the vibrant Web community, even before this pulse is materialized in the form of mature conference papers.

From the onset, the International Conference on Web Engineering (ICWE) has been conscious of the important role played by workshops in Web Engineering. The 2015 edition is no exception. We were specifically looking for topics at the boundaries of Web Engineering, aware that it is by pushing the borders that science and technology advance. The result was three workshops that were successfully held in Rotterdam on June 23, 2015:

- *NLPIT 2015*: First International Workshop on Natural Language Processing for Informal Text
- *PEWET 2015*: First Workshop on PErvasive WEb Technologies, trends and challenges
- *SoWeMine 2015*: First International Workshop in Mining the Social Web

The workshops accounted for up to 69 participants and 17 presentations, which included two keynotes, namely:

- "Hacking a Way Through the Twitter Language Jungle: Syntactic Annotation, Tagging, and Parsing of English Tweets" by Nathan Schneider
- "Fractally-Organized Connectionist Networks: Conjectures and Preliminary Results" by Vincenzo De Florio

As an acknowledgment of the quality of the workshop program, we are proud that we could reach an agreement with Springer for the publication of all accepted papers in Springer's Lecture Notes in Computer Science (LNCS) series. We opted for post-workshop proceedings, a publication modality that allowed the authors – when preparing the final version of their papers for inclusion in the proceedings – to take into account the feedback they received during the workshops and to further improve the quality of their papers.

In addition to the three workshops printed in this volume, ICWE 2015 also hosted the first edition of the *Rapid Mashup Challenge*, an event that aimed to bring together researchers and practitioners specifically working on mashup tools and/or platforms. The competition was to showcase – within the strict time limit of 10 minutes – how to develop a mashup using one's own approach. The proceedings of the challenge will be printed independently.

Without enthusiastic and committed authors and organizers, assembling such a rich workshop program and this volume would not have been possible. Thus, our first thanks go to the researchers, practitioners, and PhD students who contributed to this volume with their works. We thank the organizers of the workshops who reliably managed the organization of their events, the selection of the highest-quality papers, and the moderation of their events during the workshop day. Finally, we would like to thank the General Chair and Vice-General Chair of ICWE 2015, Flavius Frasincar and Geert-Jan Houben, respectively, for their support and trust in our work. We enjoyed organizing this edition of the workshop program, reading the articles, and assembling the post-workshop proceedings in conjunction with the workshop organizers. We hope you enjoy in the same way the reading of this volume.

July 2015   Florian Daniel
                                                                                              Oscar Diaz

Preface

The preface of this volume collects the prefaces of the post-workshop proceedings of the individual workshops. The actual workshop papers, grouped by event, can be found in the body of this volume.

First International Workshop on Natural Language Processing for Informal Text (NLPIT 2015)

Organizers: Mena B. Habib, University of Twente, The Netherlands; Florian Kunneman, Radboud University, The Netherlands; Maurice van Keulen, University of Twente, The Netherlands

The rapid growth of Internet usage in the last two decades adds new challenges to understanding the informal user generated content (UGC) on the Internet. Textual UGC refers to textual posts on social media, blogs, emails, chat conversations, instant messages, forums, reviews, or advertisements that are created by end-users of an online system. A large portion of language used on textual UGC is informal. Informal text is the style of writing that disregards language grammars and uses a mixture of abbreviations and context dependent terms. The straightforward application of state-of-the-art Natural Language Processing approaches on informal text typically results in a significantly degraded performance due to the following reasons: the lack of sentence structure; the lack of enough context required; the uncommon entities involved; the noisy sparse contents of users' contributions; and the untrusted facts contained.

This was the reason for organizing this workshop on Natural Language Processing for Informal Text (NLPIT) through which we hope to bring the opportunities and challenges involved in informal text processing to the attention of researchers. In particular, we are interested in discussing informal text modelling, normalization, mining, and understanding in addition to various application areas in which UGC is involved. The first NLPIT workshop was held in conjunction with ICWE: the International Conference on Web Engineering held in Rotterdam, The Netherlands, July 23–26, 2015. It was organized by Mena B. Habib and Maurice van Keulen from the University of Twente, and Florian Kunneman from Radboud University, The Netherlands.

The workshop started with a keynote presentation from Nathan Schneider from the University of Edinburgh entitled "Hacking a Way Through the Twitter Language Jungle: Syntactic Annotation, Tagging, and Parsing of English Tweets." Nathan explained how rich information structures can be extracted from informal text and represented in annotations. Tweets, informal text in general, is in a sense street language, but even street language is almost never entirely ungrammatical. So, even grammatical clues can be extracted, represented in annotations, and used to grasp the meaning of the text. We thank the Centre for Telematics and Information Technology (CTIT) for sponsoring this keynote presentation.

The keynote was followed by 4 research presentations selected from 7 submissions that NLPIT attracted. The common theme of these presentations was Natural Language Processing techniques for a multitude of languages. Among the 4 presentations, we saw Japanese, Tunisian, Kazakh, and Spanish. The first presentation was about extracting ASCII art embedded in English and Japanese texts. The second and fourth presentations were about constructing annotated corpora for use in research for the Tunesian dialect and Spanish, respectively. The third presentation was about word alignment issues in translating between Kazakh and English.

We thank all speakers and the audience for an interesting workshop with fruitful discussions. We furthermore hope that this workshop is the first of a series of NLPIT workshops.

July 2015

Mena Badieh Habib
Florian Kunneman
Maurice van Keulen

Program Committee

Alexandra Balahur	The European Commission's Joint Research Centre (JRC), Italy
Barbara Plank	University of Copenhagen, Denmark
Diana Maynard	University of Sheffield, UK
Djoerd Hiemstra	University of Twente, The Netherlands
Kevin Gimpel	Toyota Technological Institute, USA
Leon Derczynski	University of Sheffield, UK
Marieke van Erp	VU University Amsterdam, The Netherlands
Natalia Konstantinova	University of Wolverhampton, UK
Robert Remus	Universität Leipzig, Germany
Wang Ling	Carnegie Mellon University, USA
Wouter Weerkamp	904Labs, The Netherlands
Zhemin Zhu	University of Twente, The Netherlands

First Workshop on PErvasive WEb Technologies, Trends and Challenges (PEWET 2015)

Organizers: Fernando Ferri, Patrizia Grifoni, Alessia D'Andrea, and Tiziana Guzzo, Istituto di Ricerche sulla Popolazione e le Politiche Sociali (IRPPS), National Research Council, Italy

Pervasive Information Technologies, such as mobile devices, social media, cloud, etc., are increasingly enabling people to easily communicate and to share information and services by means of read-write Web and user generated contents. They influence the way individuals communicate, collaborate, learn, and build relationships. The enormous potential of Pervasive Information Technologies have led scientific communities in different disciplines, from computer science to social science, communication science, and economics, to analyze, study, and provide new theories, models, methods, and case studies. The scientific community is very interested in discussing and developing theories, methods, models, and tools for Pervasive Information Technologies. Challenging activities that have been conducted in Pervasive Information Technologies include social media management tools & platforms, community management strategies, Web applications and services, social structure and community modeling, etc.

To discuss such research topics, the PErvasive WEb Technologies, trends and challenges (PEWET) workshop was organized in conjunction with the 15th International Conference on Web Engineering - ICWE 2015. The workshop, held in Rotterdam, the Netherlands, on June 23–26, 2015, provided a forum for the discussion of Pervasive Web Technologies theories, methods, and experiences. The workshop organizers decided to have an invited talk, and after a review process selected five papers for inclusion in the ICWE workshops proceedings. Each of these submissions was rigorously peer reviewed by at least three experts. The papers were judged according to their originality, significance to theory and practice, readability, and relevance to workshop topics. The invited talk discussed the fractally-organized connectionist networks that according to the speaker may provide a convenient means to achieve what Leibniz calls "an art of complication," namely an effective way to encapsulate complexity and practically extend the applicability of connectionism to domains such as socio-technical system modeling and design.

The selected papers address two areas: i) Internet technologies, services, and data management and, ii) Web programming, application, and pervasive services.

In the "Internet technologies, services, and data management" area, papers discuss different issues such as retrieval and content management. In the current information retrieval paradigm, the host does not use the query information for content presentation. The retrieval system does not know what happens after the user selects a retrieval result and the host also does not have access to the information which is available to the retrieval system. In the paper titled "Responding to Retrieval: A Proposal to Use Retrieval Information for Better Presentation of Website Content" the author provided a better search experience for the user through better presentation of the content based on the query, and better retrieval results, based on the feedback to the retrieval system from the host server. The retrieval system shares some information with the host server and the host server in turn provides relevant feedback to the retrieval system.

Another issue discussed at the workshop was the modeling and creation of APIs, proposed in the paper titled "Internet-Based Enterprise Innovation through a Community-Based API Builder to Manage APIs" in which an API builder is proposed as a tool for easily creating new APIs connected with existing ones from Cloud-Based Services (CBS).

The Internet of Things (IoT) is addressed in the paper titled "End-User Centered Events Detection and Management in the Internet of Things" where the authors provide the design of a Web environment developed around the concept of event, i.e., simple or complex data streams gathered from physical and social sensors that are encapsulated with contextual information (spacial, temporal, thematic).

In the area "Web programming, application, and pervasive services" papers discuss issues such as the application of asynchronous and modular programming. This issue is complex because asynchronous programming requires uncoupling of a module into two sub-modules, which are non-intuitively connected by a callback method. The separation of the module spurs the birth of another two issues: callback spaghetti and callback hell. Some proposals have been developed, but none of them fully support modular programming and expressiveness without adding a significant complexity. In the paper titled "Proposals for Modular Asynchronous Web Programming: Issues & Challenges" the authors compare and evaluate these proposals, applying them to a non-trivial open source application development.

Another issue is that of "future studies," referring to studies based on the prediction and analysis of future horizons. The paper titled "Perspectives and Methods in the Development of Technological Tools for Supporting Future Studies in Science and Technology" gives a review of widely adopted approaches in future study activities, with three levels of detail. The first one addresses a wide scale mapping of related disciplines, the second level focuses on traditionally adopted methodologies, and the third one goes into greater detail. The paper also proposes an architecture for an extensible and modular support platform able to offer and integrate tools and functionalities oriented toward the harmonization of aspects related to semantics, document warehousing, and social media aspects. The success of the PEWET workshop would not have been possible without the contribution of the ICWE 2015 organizers and the workshop chairs, Florian Daniel and Oscar Diaz, the PC members, and the authors of the papers, all of whom we would like to sincerely thank.

July 2015

Fernando Ferri
Patrizia Grifoni
Alessia D'Andrea
Tiziana Guzzo

Program Committee

Ahmed Abbasi	University of Virginia, USA
Maria Chiara Caschera	CNR, Italy
Maria De Marsico	University of Rome "La Sapienza", Italy
Arianna D'Ulizia	CNR, Italy

Rajkumar Kannan   Bishop Heber College, India
Marco Padula   National Research Council, Italy
Patrick Paroubek   LIMSI-CNRS, France
Adam   Poznań University of Technology, Poland
  Wojciechowski

First International Workshop in Mining the Social Web (SoWeMine 2015)

Organizers: Spiros Sirmakessis, Technological Institution of Western Greece, Greece; Maria Rigou, University of Patras, Greece; Evanthia Faliagka, Technological Institution of Western Greece, Greece

The rapid development of modern information and communication technologies (ICTs) in the past few years and their introduction into people's daily lives has greatly increased the amount of information available at all levels of their social environment.

People have been steadily turning to the social web for social interaction, news and content consumption, networking, and job seeking. As a result, vast amounts of user information are populating the social Web. In light of these developments the social mining workshop aims to study new and innovative techniques and methodologies on social data mining.

Social mining is a relatively new and fast-growing research area, which includes various tasks such as recommendations, personalization, e-recruitment, opinion mining, sentiment analysis, searching for multimedia data (images, video, etc.).

This workshop aimed to study (and even go beyond) the state of the art on social web mining, a field that merges the topics of social network applications and web mining, which are both major topics of interest for ICWE. The basic scope is to create a forum for professionals and researchers in the fields of personalization, search, text mining, etc. to discuss the application of their techniques and methodologies in this new and very promising research area.

The workshop tried to encourage the discussion on new emergent issues related to current trends derived from the creation and use of modern Web applications.

Six very interesting presentations took place in two sessions

- *Session 1: Information and Knowledge Mining in the Social Web*

  - "Sensing Airport Traffic by Mining Location Sharing Social Services" by John Garofalakis, Ioannis Georgoulas, Andreas Komninos, Periklis Ntentopoulos, and Athanasios Plessas, University of Patras, Greece & University of Strathclyde, Glasgow, UK

    The paper works with location sharing social services; quite popular among mobile users resulting in a huge social dataset. Authors consider location sharing social services' APIs endpoints as "social sensors" that provide data revealing real-world interactions. They focus on check-ins at airports performing two experiments: one analyzing check-in data collected exclusively from Foursquare and another collecting additional check-in data from Facebook. They compare the two location sharing social platforms' check-ins and show in Foursquare that data can be indicative of the passengers' traffic, while their number is hundreds of times lower than the number of actual traffic observations.

  - "An Approach for Mining Social Patterns in the Conceptual Schema of CMS-based Web Applications" by Vassiliki Gkantouna, Athanasios Tsakalidis, Giannis Tzimas, and Emmanouil Viennas, University of Patras, & Technological Educational Institute of Western Greece, Greece

In this work, authors focus on CMS-based web applications that exploit social networking features and propose a model-driven approach to evaluating their hypertext schema in terms of the incorporated design fragments that perform a social network related functionality. Authors have developed a methodology, which, based on the identification and evaluation of design reuse, detects a set of recurrent design solutions denoting either design inconsistencies or effective reusable social design structures that can be used as building blocks for implementing certain social behavior in future designs.

- "An E-recruitment System Exploiting Candidates' Social Presence" by Evanthia Faliagka, Maria Rigou, and Spiros Sirmakessis, Technological Educational Institution of Western Greece, University of Patras, & Hellenic Open University, Greece

    This work aims to help HR Departments in their job. Applicant personality is a crucial criterion in many job positions. Choosing applicants whose personality traits are compatible with job positions is the key issue for HR. The rapid deployment of social web services has made candidates' social activity much more transparent, giving us the opportunity to infer features of candidate personality with web mining techniques. In this work, a novel approach is proposed and evaluated for automatically extracting candidates' personality traits based on their social media use.

- *Session 2: Mining the Tweets*

    - "#nowplaying on #Spotify: Leveraging Spotify Information on Twitter for Artist Recommendations" by Martin Pichl, Eva Zangerle, and Günther Specht, Institute of Computer Science, University of Innsbruck, Austria

        The rise of the Web has openned new distribution channels like online stores and streaming platforms, offering a vast amount of different products. To help customers find products according to their taste on those platforms, recommender systems play an important role. Authors present a music recommendation system exploiting a dataset containing listening histories of users, who posted what they are listening to at the moment on Twitter. As this dataset is updated daily, they propose a genetic algorithm, which allows the recommender system to adopt its input parameters to the extended dataset.

    - "Retrieving Relevant and Interesting Tweets during Live Television Broadcasts" by Rianne Kaptein, Yi Zhu, Gijs Koot, Judith Redi, and Omar Niamut, TNO, The Hague & Delft University of Technology, The Netherlands

        The use of social TV applications to enhance the experience of live event broadcasts has become an increasingly common practice. An event profile, defined as a set of keywords relevant to an event, can help to track messages related to these events on social networks. Authors propose an event profiler that retrieves relevant and interesting tweets in a continuous stream of event-related tweets as they are posted. For testing the application they have executed a user study. Feedback is collected during a live broadcast by giving the participant the option to like or dislike a tweet, and by judging a selection of tweets on relevancy and interest in a post-experiment questionnaire.

- "Topic Detection in Twitter Using Topology Data Analysis" by Pablo Torres-Tramon, Hugo Hromic, and Bahareh Heravi, Insight Centre for Data Analytics, National University of Ireland, Galway

  The authors present automated topic detection in huge datasets in social media. Most of these approaches are based on document clustering and burst detection. These approaches normally represent textual features in standard n-dimensional Euclidean metric spaces. Authors propose a topic detection method based on Topology Data Analysis that transforms the Euclidean feature space into a topological space where the shapes of noisy irrelevant documents are much easier to distinguish from topically-relevant documents.

July 2015

Spiros Sirmakessis
Maria Rigou
Evanthia Faliagka

Program Committee

Olfa Nasraoui	University of Louisville, USA
Martin Rajman	EPFL, Switzerland
Evanthia Faliagka	Technological Institution of Western Greece
John Garofalakis	University of Patras, Greece
Maria Rigkou	University of Patras, Greece
Spiros Sioutas	Ionian University, Greece
Spiros Sirmakessis	Technological Educational Institution of Western Greece
John Tsaknakis	Technological Educational Institution of Western Greece
John Tzimas	Technological Educational Institution of Western Greece
Vasilios Verikios	Hellenic Open University, Greece

Contents

First International Workshop in Mining the Social Web (SoWeMine 2015)

First International Workshop on Natural Language Processing for Informal Text (NLPIT 2015)

Constructing Linguistic Resources for the Tunisian Dialect Using Textual User-Generated Contents on the Social Web

Jihen Younes[1(✉)], Hadhemi Achour[2], and Emna Souissi[1] ·

[1] Université de Tunis, ENSIT, 1008 Montfleury, Tunisia
`jihene.younes@gmail.com, emna.souissi@esstt.rnu.tn`
[2] Université de Tunis, ISGT, LR99ES04 BESTMOD, 2000 Le Bardo, Tunisia
`Hadhemi_Achour@yahoo.fr`

Abstract. In Arab countries, the dialect is daily gaining ground in the social interaction on the web and swiftly adapting to globalization. Strengthening the relationship of its practitioners with the outside world and facilitating their social exchanges, the dialect encompasses every day new transcriptions that arouse the curiosity of researchers in the NLP community. In this article, we focus specifically on the Tunisian dialect processing. Our goal is to build corpora and dictionaries allowing us to begin our study of this language and to identify its specificities. As a first step, we extract textual user-generated contents on the social Web, we then conduct an automatic content filtering and classification, leaving only the texts containing Tunisian dialect. Finally, we present some of its salient features from the built corpora.

Keywords: Tunisian dialect · Language identification · Corpus construction · Dictionary construction · Social web textual contents

1 Introduction

The Arabic language is characterized by its plurality. It consists of a wide variety of languages, which include the modern standard Arabic (MSA), and a set of various dialects differing according to regions and countries. The MSA is one of the written forms of Arabic that is standardized and represents the official language of Arab countries. It is the written form generally used in press, media, official documents, and that is taught in schools. Dialects are regional variations that represent naturally spoken languages by Arab populations. They are largely influenced by the local historical and cultural specificities of the Arab countries [1]. They can be very different from each other and also present significant dissimilarities with the MSA.

While many efforts have been undertaken during the last two decades for the automatic processing of MSA, the interest in processing dialects is quite recent and related works are relatively few. Most of the Arabic dialects are today under-resourced languages and some of them are unresourced. Our work is part of the contributions to automatic processing of the Tunisian dialect (TD). The latter faces a

© Springer International Publishing Switzerland 2015
F. Daniel and O. Diaz (Eds.): ICWE 2015 Workshops, LNCS 9396, pp. 3–14, 2015.
DOI: 10.1007/978-3-319-24800-4_1

major difficulty which is the almost total absence of resources (corpora and lexica), useful for developing TD processing tools such as morphological analyzers, POS taggers, information extraction tools, etc.

As Arabic materials are written essentially in MSA, we propose in this work to exploit informal textual content generated by Tunisian users on the Internet, particularly their exchanges on social networks, for harvesting texts in TD and building TD language resources. Indeed, social exchanges have undergone a swift evolution with the emergence of new communication tools such as SMS, fora, blogs, social networks, etc. This evolution gave rise to a recent form of written communication namely the electronic language or the network language. In Tunisia, this language appeared with SMS in the year 2000 with the emergence of mobile phones. Users began to create their own language by using the Tunisian dialect and by enriching it with words of different origins. According to latest figures (December, 2014) from the Internet World Stats[1], the number of Internet users in Tunisia reached 5,408,240 (49% of the population), giving the Tunisian electronic language free field to be further diversified and enriched in other contexts namely blogs, fora and social websites.

Starting from these informal data, mainly provided in our case by social networks contents, we propose in this paper to extend our previous work [4], in which we collected a corpus of written TD messages in Latin transcription (TLD), by proposing an enhanced approach for also automatically identifying TD messages in Arabic transcription (TAD), in order to build a richer set of TD language resources[2] (corpora and lexica).

In what follows, related work is presented in Section 2. Section 3 is devoted to the construction of TD language resources. In this section, we first expose difficulties of collecting TD messages. We will then present the different steps of the adopted approach for extracting and identifying TD words and messages. A brief overview in figures, on the salient features of the obtained corpora (TAD corpus and TLD corpus) is presented in Section 4. Results obtained in an evaluation of the proposed approach for identifying TD language will be discussed in Section 5.

2   Related Work

While reviewing the literature on available language resources related to Arabic dialects, we quickly notice that there is little written material in the Tunisian dialect. To the best of our knowledge, it is since 2013 that work dealing with the automatic processing of TD language and building the required linguistic resources has begun to be published.

As the most used written form of Arabic is MSA, almost all Arabic linguistic resources content is essentially in MSA. In order to address the lack of data in Arabic dialects, some researchers have explored the idea of using existing MSA resources to automatically generate the equivalent dialectal resources. This is for instance, the case of Boujelbane et al. [2], who proposed an automatic corpus generation in the Tunisian dialect, from the Arabic Tree bank Corpus [3]. Their approach relies on a set of

---

[1]   http://www.internetworldstats.com/africa.htm#tn
[2]   These resources may be obtained by contacting the first author.

transformation rules and a bilingual lexicon MSA versus TD language. Note however that in [2], Boujelbane et al. have considered only the transformation of verbal forms.

In our previous work [4], we focused on the Latin transcription of the Tunisian dialect and built a TD corpus written in Latin alphabet, composed of 43 222 messages. Multiple data sources were considered including written messages sent from mobile phones, Tunisian fora and websites, and mainly Facebook network.

Work related to other Maghrebi dialects may be cited such as those concerned with the Algerian and Moroccan dialects: Meftouh et al. [5] aim to build an MSA-Algerian Dialects translation system. They started from scratch and manually built a set of linguistic resources for an Algerian dialect (specific to Annaba region): a corpus of manually transcribed texts from speech recordings, a bilingual lexicon (MSA-Annaba Dialect) and a parallel corpus also constructed by hand. In [6], an Algerian Arabic-French code-switched corpus was collected by crawling an Algerian newspaper website. It is composed of 339 504 comments written in Latin alphabet. MDED presented in [7] is a bilingual dictionary MSA versus a Moroccan dialect. It counts 18 000 entries, mainly constructed by manually translating an MSA dictionary and a Moroccan dialect dictionary.

As for non Maghrebi dialects, there are several dialectal Arabic resources we can mention such as YADAC corpus presented in [8] by Al-Sabbagh and Girju, that is compiled using Web data from microblogs, blogs/fora and online knowledge market services. It focused on Egyptian dialect which was identified, mainly using Egyptian function words specific to this dialect. Diab et al. [9], Elfardy and Diab [10], worked on building resources for Egyptian, Iraqi and Levantine dialects and built corpora, mainly from blogs and forum messages. Further work on the identification of Arabic dialects was conducted by Zaidan and Callison-Burch [11, 12], who built an Arabic commentary dataset rich in dialectal content from Arabic online newspapers. Cotterell and Callison-Burch [13] dealt with several Arabic dialects and collected data from newspaper websites for user commentary and Twitter. They built a multi-dialect, multi-genre, human annotated corpus of Levantine, Gulf, Egyptian, Iraqi and Maghrebi dialects. In [13], classification of dialects is carried out using machine learning techniques (Naïve Bayes and Support Vector Machines), given a manually annotated training set.

In the aim of developing a system able to recognize written Arabic dialects (mainly, the two groups: Maghrebi dialects and Middle-East dialects), Saadane et al. [1] constructed, from the internet and some speech transcription applications, a corpus of dialectal texts, written in Latin Alphabet, then transliterated it in Arabic Alphabet.

3   Construction of TD Linguistic Resources

We proceeded in our construction approach to collecting linguistic productions provided by users of social websites, more particularly the Facebook social network. Our choice was based on the fact that at the present time, social networks are among the most users requested means of communication. According to Thecountries.com[3],

---

[3] http://www.thecountriesof.com/top-10-countries-with-most-facebook-users-in-the-world-2013/

Facebook, with the largest number of users, is one of the most popular social sites in 2013. Tunisians prefer Facebook over other social networks. The site StatCounter.com[4] conducted a statistical study in 2014 which showed that the use rate of Facebook in Tunisia is around 97%. YouTube monopolizes the second position (1.3%) and Twitter the third one (1.01%).

3.1   Difficulties in Collecting TD Messages

The extraction of the Tunisian dialect from informal content on the Internet is a nontrivial task. Tunisian electronic language is in fact, an interference between the TD and the network language. It is basically a fusion with other languages (French, English, etc.), with a margin of individualization, giving the user the freedom to write without depending on spelling constraints or grammar rules. This margin of freedom increases the number of possible transcriptions for a given word, and reveals in return a considerable challenge in the treatment of this new form of writing. As for its writing system, it can vary from Latin to Arabic. Looking at the social web pages, it seems clear that Tunisians are more likely to transcribe their messages with Latin letters. The lack of Arabic keyboards in the beginning of web and mobile era reinforced this preference, not to mention the factors of linguistic fusion of written standard Arabic (MSA) and the neighboring languages, as well as the influence of colonization, migration, and the neo-cultures.

Whether for written TD with the Latin or the Arabic alphabet, multilingualism is one of the most observed phenomena. Practitioners of this form of writing can introduce words from several languages, in their standard or SMS form (textese)[5]. The message in Fig. 1 shows an example of multilingualism in the TLD and the TAD.

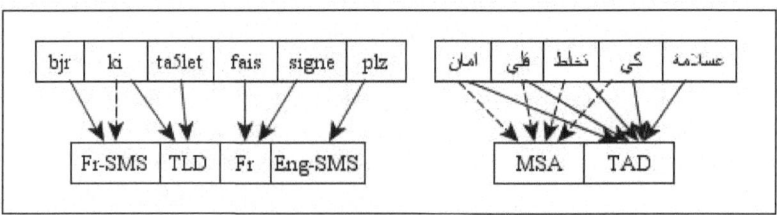

Fig. 1. Examples of TD messages [4]

The TLD message in Fig. 1 begins with the word " bjr ", a French word written in SMS language, it is the abbreviation of the word " bonjour " which means " hello ". The word " ki " means in this context " when " and " ta5let " mean " you come " in TD. The words " fais " and " signe " which, as an expression, mean " let me know " are written

---

4   http://gs.statcounter.com/#desktop-social_media-TN-monthly-201303-201403-bar
5   "form of written language as used in text messages and other digital communications, characterized by many abbreviations and typically not following standard grammar, spelling, punctuation and style". (www.dictionary.reference.com)

in standard French, and the word " plz " means " please " in English SMS. As for the TAD example, it is practically the translation of the TLD message. We notice the high rate of words that can be considered simultaneously as TAD and MSA words.

Although the multilingualism phenomenon reveals the richness of the TD, it poses, in return, a problem in the language ambiguity (Table 1).

Table 1. Examples of ambiguous word in TD

Word	خاطر		Bard		Flous	
Meaning	TAD	MSA	TLD	English	TLD	French
	Because	*spirit*	*cold*	*Poet*	*Money*	*fuzzy*

This language ambiguity complicates the process of automatic corpus building for TD. The difficulty lies in the automatic classification of extracted messages and in the decision to make if they contain ambiguous words. That is to say, how can we classify them into TD messages and non TD messages?

The adopted approach, presented in the next section, is quite straightforward and is mainly based on the detection of TD unambiguous words using pre-built TD lexica for identifying TD messages. This approach is a starting solution to accumulate an initial amount of resources that we can use later to implement and test machine learning techniques.

3.2   TD Lexicon Construction

In the first step of our study, we focused on building lexica for the TAD and the TLD. Work, rather manual, was performed, consisting in selecting personal messages, comments and posts from social sites. Thus, a corpus of 6 079 messages written in TLD was built. This corpus allowed us to identify, after cleaning punctuation and foreign words, a lexicon of 19 763 TLD words. We manually assigned to each word, its potential transliterations in Arabic alphabet (example: tounes ↔ تونس) in order to get a set of TAD words.

A reverse dictionary was automatically generated through the TLD→TAD inputs, consisting of 18 153 entries. This TAD→TLD dictionary associates each word written in Arabic letters its set of transliterations written in Latin letters (Table 2).

Table 2. Sample entries in the TD dictionaries

Dictionary	Number of inputs	Example
TLD→TAD	19 763	سَاحَة I صَحَّة I Sa7a
TAD→TLD	18 153	صَحَّة I saha I sa7a I sahha I sa77a

3.3   Message Extraction

In the message extraction, a tool that allows us to return the comments of a Tunisian page through its unique identifier on Facebook was developed. Different types of pages were exploited to ensure the diversity of the corpus (media, politics, sports, etc.) and cover the maximum of the vocabulary used.

The messages we need for the corpus should be written in TD. However, the automatic retrieval returned 73 024 messages consisting of links, advertisements (spam), messages written in Arabic letters (MSA or TD) and messages written in other languages (French, English, French-SMS, etc.). Therefore, we developed a filtering and classification tool that detects the type of each message and classifies it as TD or non TD. To do this, we used the built lexica TLD and TAD, as well as other lexica for MSA, French, French-SMS, English, and English-SMS (Table 3).

Table 3. Lexica used in the filtering steps

Lexicon	Number of inputs	Writing system
TLD	19 763	Latin
TAD	18 153	Arabic
MSA	449 801	Arabic
Fr	336 531	Latin
Fr-SMS	770	Latin
Eng	354 986	Latin
Eng-SMS	950	Latin

3.4 Filtering and Classification

Our filtering and classification approach is based primarily on the lexica. To perform automatic filtering, three steps were followed (Fig. 2):

- First filter: cleaning the messages of advertisements and spam. This step is mainly based on web links detection and returned a total of 66 098 user comments.
- Second filter: filtering and dividing the messages in two categories (Arabic alphabet or Latin alphabet). At the end of this filtering, we find that more than 72% of extracted messages are written in Latin characters, which confirms the idea that we advanced in Section 3.1 on the preferences of Tunisians in the transcription of their messages on the social web.
- Third filter (classification): classifying the messages according to their language (TD or non TD). Since the collected messages usually contain several ambiguous words, we tried to identify, using lexica of Table 3, the language of each word in a message and consider only the unambiguous TD words (belonging only to TD lexica). A message is thus, identified as a TD message, only if it contains at least one unambiguous TD word. Table 4 shows an example of the word identification in the classification step.

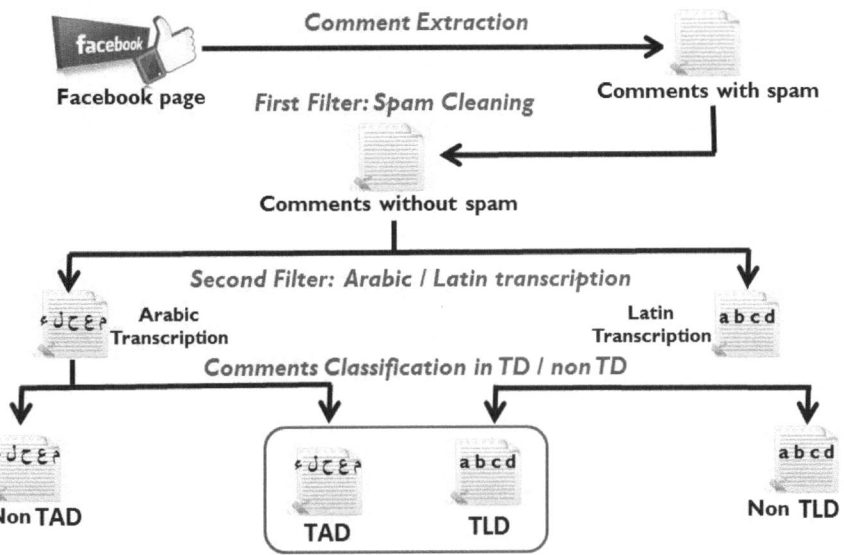

Fig. 2. Automatic filtering steps

Table 4. Word identification in the classification step

Dialect type	TLD			TAD		
Message	bjr, ki ta5let 9oli			عسلامة، كي تخلط قلي		
	Word	**Ambiguity**	**Language**	**Word**	**Ambiguity**	**Language**
	Bjr		Fr-SMS	عسلامة		**TAD**
Identified words	Ki	✓	TLD Fr-SMS	كي	✓	TAD MSA
	ta5let		**TLD**	تخلط	✓	TAD MSA
	9oli		**TLD**	قلي	✓	TAD MSA

The messages shown in Table 4 are considered in TD language, as they contain unambiguous dialect words (" ta5let " and " 9oli " in TLD and " عسلامة " in TAD). The classification protocol is summarized in Fig. 3.

Finally, and after the automatic classification step, we obtained a TLD corpus consisting of 31 158 messages, and a TAD corpus consisting of 7 145 messages (Fig. 4).

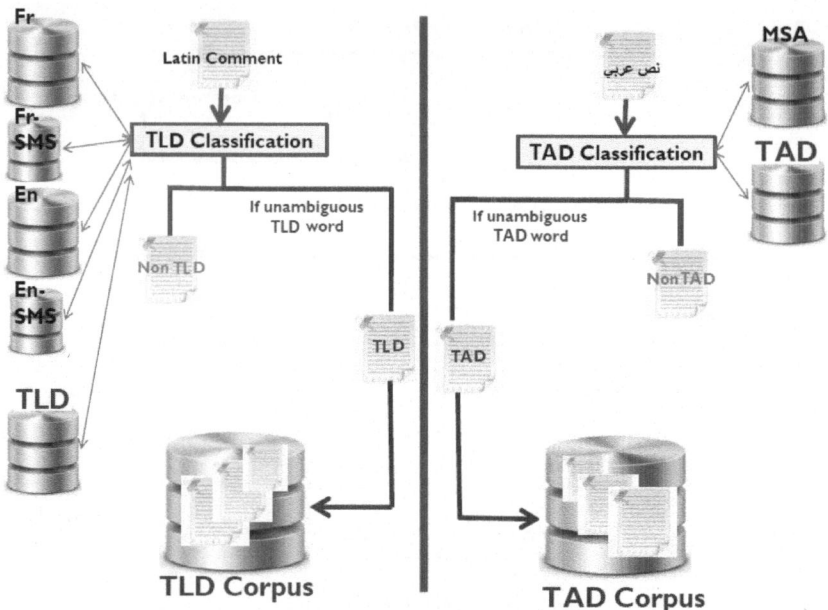

Fig. 3. Classification step

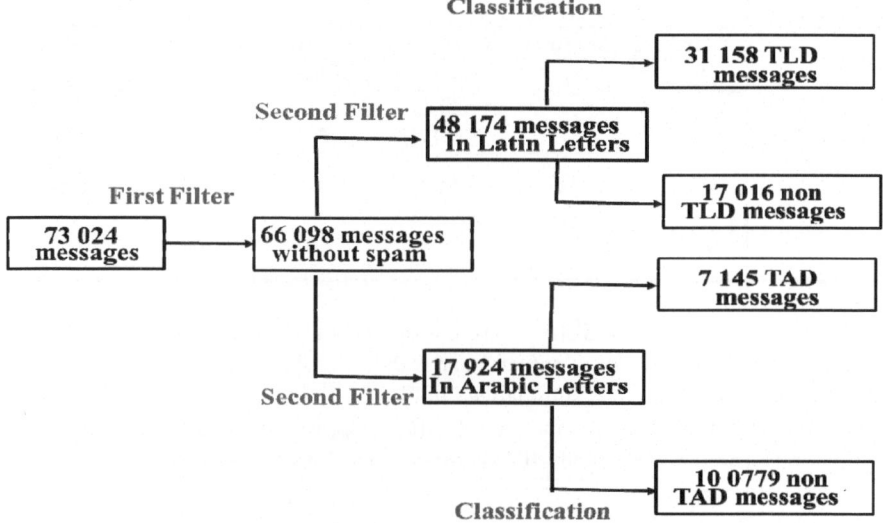

Fig. 4. Results of the filtering and the classification steps

4   Characteristics of the Corpora

We present in what follows, a brief study on some features of both TLD and TAD obtained corpora consisting respectively of 420 897 and 160 418 words.

1. *Message sizes.* In TLD, the shortest message consists of a single word and 2 characters. The longest consists of 307 words and 1642 characters. The messages are longer in TAD, the maximum size is 464 words and 2589 characters.
2. *Word sizes.* On average, a word in the TD corpora consists of 5 characters. In the lexicon of the TLD corpus, the average size is 7 characters, and in the lexicon of the TAD corpus, the average is equal to 6 characters.
3. *Multilingualism.* In the spoken Tunisian, more than three different languages can be found in a single sentence, the most common are: TD, French and English. As for the written, it is much more complex, a TD word can be written in several ways. There are no specific rules as it is not an official or a taught language, but it is Tunisians' mother tongue. According to the counting made on the corpus, several intersections were identified between the TD and other languages. We noticed indeed, the large number of words in common between TLD and English, and between TAD and MSA. Fig. 5 shows this overlapping and gives the percentage of words in the TD corpora, which are in common with other languages.

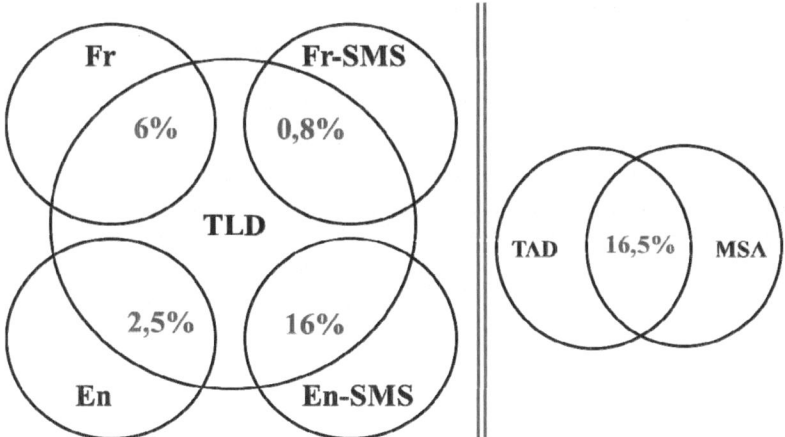

Fig. 5. Overlapping between TD and other languages in the TD Corpora

Regarding the overlapping between TAD and MSA languages, we can notice that ambiguous (common) words can be of mainly three types:

- *True cognates.* These are words which are written in the same way in TAD and in MSA and have also the same meaning. Example: "ناس" is a MSA and a TAD word meaning " people ".

- *Homographs and homophones* are words having the same written form in TAD and in MSA, but have a different meaning, as for example the common word " خَاطِر " which mean " spirit " in MSA and " because " in TAD.
- *Words with ambiguous vocalization.* This kind of words have the same written form and the same meaning, but have different vocalization and thus different pronunciations. For example the TAD word " خْرَجَ " and the MSA one " خَرَجَ " have both the same meaning " He got out ". As practitioners of TAD tend to overlook the vowels, it is difficult to determine the language of non vocalized words. This problem does not occur in TLD given the frequent use of the letters " a ", " i ", " o ", the respective equivalents of Arabic vowels " Fatha "(́), " Kasra " (̣) and " Dhamma " (̣).

4. *Word frequencies.* After extracting the word frequencies, we noticed that the most common TAD words are ambiguous function words (" و ", " في ", " على " ...), since they are shared between the two entities TAD and MSA. Therefore, we cannot base our classification approach on the function words recognition. Regarding the TLD, we noticed, among the most frequent words, the presence of unambiguous particles (" fi ", " ya ", " el ", " bech ", etc.) which are not used in French and English. Consequently, these words can help us identify the language of the messages and improve our classification approach.

5. *Word stretching* (Repeated sequence of letters). This phenomenon consists in repeating a character several times to emphasize the intensity of emotion. This phenomenon is encountered in both TLD and TAD corpora (" Baaaarcha " ↔ " باااارشا " which means " much "). In the TLD corpus, 8839 (2%) words contain a repeated sequence of letters. This number decreases to 1615(1%) in the TAD corpus.

6. *Use of digits.* This phenomenon concerns only the TLD. Arabic letters that have no equivalent letter in the Latin Aphabet, are replaced by digits. Table 5 presents some equivalents.

Table 5. Equivalences between digits and Arabic letters

Arabic letter	أ	ح	خ	ع	غ	ق
Equivalent digit	2	7	5	3	8	9

5   Evaluation

In order to assess the ability of the adopted classification approach to correctly separate TD from non TD messages, we conducted an evaluation on a portion of each corpus. We extracted 10% of messages from each of the raw Arabic and Latin corpora, we then proceeded to manually verify the results of their automatic classification (Table 6).

Table 6. Results of the automatic classification

Test Corpus	Number of messages	Automatic classification			
		TD	True TD	Non TD	True non TD
Latin	4 817	2 901	2 888	1 916	1 493
Arabic	1 792	708	669	1 084	645

Evaluation results, calculated in terms of accuracy, precision, recall and F-measure are given in Table 7 below.

Table 7. Evaluation results

Test corpus	Accuracy	Precision	Recall	F-measure
Latin corpus	0,9	0,99	0,87	0,92
Arabic corpus	0,73	0,94	0,6	0,73

As shown in Table 7, precision is high for both Latin TD (0.99) and Arabic TD (0.94). In fact, the method we used is very selective in its choice of TD messages, since it is based on the TD unambiguous words. Rare cases of messages that were incorrectly classified as TD, correspond to Arabic messages containing words that the system considered unambiguous TAD, because they do not belong to the MSA lexicon, when in reality they are shared between both TAD and MSA languages. As for the corpus written in Latin letters, these cases correspond to some messages that contain Tunisian proper nouns belonging only to the TLD lexicon, but the rest of its words are non TLD.

In terms of accuracy and recall, we note that results for the TLD corpus are better than those achieved for the TAD corpus. Indeed, we found a relatively high intersection between the TAD and the MSA. Unrecognized TD messages may contain several dialectal words, but our system is unable to classify them as TD since these words are also potential MSA words. The major limitation of the lexicon based approach, is the fact that we are considering each word separately without taking into account their global context.

6   Conclusion

We presented in this paper an approach for automatically extracting and identifying TD messages from web social textual contents, in order to build TD corpora. We also, exposed the salient features of their Arabic and Latin forms. Due to the lack of language resources for the Tunisian dialect, we started by a quite simple classification approach, mainly based on detecting non ambiguous TD words using TD lexica. Our goal was to build initial TD corpora that would be a starting point to begin working on the automatic processing of this language. The proposed approach reveals a crucial problem which is the language ambiguity of words composing a message, mainly due to the overlapping between Arabic TD and MSA on the one hand, and between Latin

TD and French and English languages on the other hand. This phenomenon arises with greater extent between TAD and MSA and led to less efficient results for the Arabic TD identification, with an accuracy of 73% for TAD against 90% for TLD. In order to enhance this work, we are planning to use the collected resources to implement and test additional language identification approaches, especially classification approaches based on machine learning techniques. We also aim to move to the annotation of the built corpora and the development of various TD NLP tools, mainly TD POS tagging and parsing tools.

References

1. Saadane, H., Guidere, M., Fluhr, C.: La reconnaissance automatique des dialectes arabes à l'écrit. In: Colloque International Traduction et Champs Connexes, Quelle Place Pour La Langue Arabe Aujourd'hui?, pp. 18–20, Alger (2013)
2. Boujelbane, R., Khemekhem, M., Belguith, L.: Mapping rules for building a Tunisian dialect lexicon and generating corpora. In: International Joint Conference on Natural Language Processing, pp. 419–428, Nagoya (2013)
3. Maamouri, M., Bies, A.: Developing an Arabic treebank: methods, guidelines, procedures, and tools. In: Workshop on Computational Approaches to Arabic Script-based Languages, Geneva (2004)
4. Younes, J., Souissi, E.: A quantitative view of Tunisian dialect electronic writing. In: 5th International Conference on Arabic Language Processing, pp. 63–72, Oujda (2014)
5. Meftouh, K., Bouchemal, N., Smaïli, K.: A study of a non-resourced language: an Algerian dialect. In: 3rd International Workshop on Spoken Languages Technologies for Under-resourced Languages, Cape Town (2012)
6. Cotterell, R., Renduchintala, A., Saphra, N., Callison-Burch, C.: An Algerian Arabic-French code-switched corpus. In: 9th International Conference on Language Resources and Evaluation, Reykjavik (2014)
7. Tachicart, R., Bouzoubaa, K., Jaafar, H.: Building a Moroccan dialect electronic dictionary (MDED). In: 5th International Conference on Arabic Language Processing, pp. 216–221, Oujda (2014)
8. Al-Sabbagh, R., Girju, R.: Yet another dialectal Arabic corpus. In: 8th International Conference on Language Resources and Evaluation, pp. 2882–2889, Istanbul (2012)
9. Diab, M., Habash, N., Rambow, O., Altantawy, M., Benajiba, Y.: COLABA: Arabic dialect annotation and processing. In: 7th International Conference on Language Resources and Evaluation, pp. 66–74, Valletta (2010)
10. Elfarady, H., Diab, M.: Simplified guidelines for the creation of large scale dialectal Arabic annotations. In: 8th International Conference on Language Resources and Evaluation, pp. 371–378, Istanbul (2012)
11. Zaidan, O.F., Callison-Burch, C.: The Arabic online commentary dataset: an annotated dataset of informal Arabic with high dialectal content. In: Association for Computational Linguistics, pp. 37–41, Portland (2011)
12. Zaidan, O.F., Callison-Burch, C.: Arabic dialect identification. In: Association for Computational Linguistics, pp. 171–202, Baltimore (2014)
13. Cotterell, R., Callison-Burch, C.: A multi-dialect, multi-genre corpus of informal written Arabic. In: 9th International Conference on Language Resources and Evaluation, pp. 241–245, Reykjavik (2014)

Spanish Treebank Annotation of Informal Non-standard Web Text

Mariona Taulé[2]([✉]), M. Antonia Martí[2], Ann Bies[1], Montserrat Nofre[2],
Aina Garí[2], Zhiyi Song[1], Stephanie Strassel[1], and Joe Ellis[1]

[1] Linguistic Data Consortium, University of Pennsylvania,
3600 Market Street, Suite 801, Philadelphia, PA 19104, USA
[2] CLiC, University of Barcelona, Gran Via 588, 08007 Barcelona, Spain
mtaule@ub.edu

Abstract. This paper presents the Latin American Spanish Discussion Forum Treebank (LAS-DisFo). This corpus consists of 50,291 words and 2,846 sentences that are part-of-speech tagged, lemmatized and syntactically annotated with constituents and functions. We describe how it was built and the methodology followed for its annotation, the annotation scheme and criteria applied for dealing with the most problematic phenomena commonly encountered in this kind of informal unedited web text. This is the first available Latin American Spanish corpus of non-standard language that has been morphologically and syntactically annotated. It is a valuable linguistic resource that can be used for the training and evaluation of parsers and PoS taggers.

1 Introduction

In this article we present the problems found and the solutions adopted in the process of the tokenization, part-of-speech (PoS) tagging and syntactic annotation of the Latin American Spanish Discussion Forum Treebank (LAS-DisFo).[1] This corpus consists of a compilation of textual posts and includes suggestions, ideas, opinions and questions on several topics including politics and technology.

Like chats, tweets, blogs and SMS these texts constitute a new genre that is characterized by an informal, non-standard style of writing, which shares many features with spoken colloquial communication: the writing is spontaneous, performed quickly and usually unedited. At the same time, to recover the lack of

This material is based on research sponsored by Air Force Research Laboratory and Defense Advanced Research Projects Agency under agreement number FA8750-13-2-0045. The U.S. Government is authorized to reproduce and distribute reprints for Governmental purposes notwithstanding any copyright notation thereon. The views and conclusions contained herein are those of the authors and should not be interpreted as necessarily representing the official policies or endorsements, either expressed or implied, of Air Force Research Laboratory and Defense Advanced Research Projects Agency or the U.S. Government.

[1] A Discussion Forum is an online asynchronous discussion board where people can hold conversations in the form of posted messages.

© Springer International Publishing Switzerland 2015
F. Daniel and O. Diaz (Eds.): ICWE 2015 Workshops, LNCS 9396, pp. 15–27, 2015.
DOI: 10.1007/978-3-319-24800-4_2

face-to-face interactions, the texts contain pragmatic information about mood and feelings often expressed by paratextual clues: emoticons, capital letters and non-conventional spacing, among others. As a consequence, the texts produced contain many misspellings and typographic errors, a relaxation of standard rules of writing (i.e. the use of punctuation marks) and an unconventional use of graphic devices such as the use of capital letters and the repetition of some characters.

These kinds of texts are pervasive in Internet data and pose difficult challenges to Natural Language Processing (NLP) tools and applications, which are usually developed for standard and formal written language. At the same time, they constitute a rich source of information for linguistic analysis, being samples of real data from which we can acquire linguistic knowledge about how languages are used in new communication modalities. Consequently, there is an increasing interest in the analysis of informal written texts, with annotated corpora where these characteristics are explicitly tagged and recovered as one of the crucial sources of information to fill this need. In particular, this Latin American Spanish Treebank is being developed in support of DARPA's Deep Exploration and Filtering of Text (DEFT) program, which will develop automated systems to process text information and enable the understanding of connections in text that might not be readily apparent to humans. The Linguistic Data Consortium (LDC) supports the DEFT Program by collecting, creating and annotating a variety of informal data sources in multiple languages to support Smart Filtering, Relational Analysis and Anomaly Analysis.

This paper is structured as follows. After a brief introduction to the related work (section 2), we present how the LAS-DisFo was built (section 3). Then, we describe the annotation process carried out (section 4), followed by the annotation scheme and criteria adopted (section 5). First, we focus on the word-level tokenization and morphological annotation (subsection 5.1) and, then, on the sentence segmentation (subsection 5.2) and syntactic annotation (subsection 5.3). Final remarks are presented in (section 6).

2   Related Work

It is well known that NLP tools trained on well-edited texts perform badly when applied to unedited web texts [7]. One of the reasons for this difficulty is the result of a mismatch between the training data, which is typically the Wall Street Journal portion of the PennTreeBank [11] in the case of English, and the corpus to be parsed. Experiments carried out with English texts such as those reported in [13] show that current parsers achieve an accuracy of 90% when they are limited to heavily edited domains, but when applied to unedited texts their performance falls to 80%, and even PoS tagging scores only slightly higher than 90%. The problem increases with morphologically rich languages such as French [14] and Spanish.

Considering that many NLP applications such as Machine Translation, Sentiment Analysis and Information Extraction need to handle unedited texts, there is a need for new linguistic resources such as annotated web text corpora to extend already existing parsers and for the development of new tools.

The annotation of unedited web corpora presents specific challenges, which are not covered by current annotations schemes and require specific tagsets and annotation criteria. This explains the increasing interest in the organization of workshops focusing on the annotation of informal written texts (EWBTL-2014; NLPIT-2015; LAW-Informal text-2015). There is an increasing interest in the development of annotated corpora of non-standard texts. These are usually small corpora in which the different web genres are represented or representative of one specific genre: English Web Treebank [2]; French Social Media Bank [14]; the No-Sta-D corpus of German non-standard varieties [6]; the #hardtoparse corpus of tweets [8], among others.

3   Latin American Spanish Discussion Forum Corpus

3.1   LDC Spanish DF Data Collection

Spanish discussion forum (DF) data was collected by LDC in support of the DEFT program, in order to build a corpus of informal written Spanish data that could also be annotated for a variety of tasks related to DEFT's goal of deep natural language understanding. DF threads were collected based on the results of manual data scouting by native Spanish speakers who searched the web for Spanish DF discussions according to the desired criteria, focusing on DF topics related to current events and other dynamic events. The Spanish data scouts were instructed to search for content on these topics that was interactive, informal, original (i.e., written by the post's author rather than quoted from another source), and in Spanish (with a particular focus on Latin American Spanish during the latter part of the collection). After locating an appropriate thread, scouts then submitted the URL and some simple judgments about the thread to a collection database via a web browser plugin. Discussion forums containing the manually collected threads were selected and the full forum sites were automatically harvested, using the infrastructure described in [9].

3.2   Latin American Spanish DF Data Selection and Segmentation

A subset of the collected Spanish DF data was selected by LDC for annotation, focusing on the portion that had been harvested from sites identified as containing primarily Latin American Spanish. The goal was to select a data set suitable for multiple levels of annotation, such as Treebank and Entities, Relations, and Events (ERE) [15]. Creating multiple annotations on the same data will facilitate experimentation with machine learning methods that jointly manipulate the multiple levels. Documents were selected for annotation based on the density of events, which was required for ERE. The resulting Latin American Spanish DF data set to be used for Spanish Treebank annotation consists of 50,291 words and 2,846 sentences in 60 files, each of them a thematically coherent fragment from a forum.

4   Annotation Process

The LAS-DisFo corpus is annotated with morphological and syntactic information by applying automatic and manually annotation processes. Firstly, the corpus was automatically tokenized, PoS tagged and lemmatized using tools from
the Freeling library[2] [12]. Then, a manual check of the output of these automatic
processes was carried out. At this level, a greater level of human intervention was
required than with standard written corpora. As we will observe in the annotation criteria sections, most of the problems arose from word tokenization and
word spellings rather than at the syntactic level.

LAS-DisFo was then subjected to a completely manual syntactic annotation
process. In order to guarantee the quality of the results, we first carried out the
constituent annotation followed by the annotation of syntactic functions.

The annotation team was made up of seven people: two senior researchers
with in-depth experience in corpus annotation that supervised the whole process;
one senior annotator with considerable experience in this field, who was respon-
sible for checking and approving the whole annotation task; and four undergrad-
uate students in their final year, who carried out the annotation task. One of the
students reviewed the morphology, two students annotated constituents and the
other two students annotated both constituents and functions. This organization
meant that the earlier annotations were revised at every stage of the process.
After one and a half months of training, the three syntactic annotators carried
out an interannotator agreement test using 10 files. These files were manually
compared and we discussed solutions for the inconsistencies that were found,
so as to minimize them. The initial guidelines were updated and the annotation process started. The team met once a week to discuss the problems arising
during the annotation process to resolve doubts and specific cases.

The annotations were performed using the AnCoraPipe annotation tool [1]
to facilitate the task of the annotators and to minimize the errors in the annotation process. The corpora texts annotated were XML documents with UTF-8
encoding.

5   Annotation Scheme and Criteria

Two main principles guided the whole annotation process. First, the source text
was maintained intact. The preservation of the original text is crucial, because
in this way the corpus will be a resource for deriving new tools for the analysis of informal Spanish language, as well as for the linguistic analysis of spontaneous written language. Second, we used a slightly modified version of the
annotation scheme followed for the morphological and syntactic tagging of the
Spanish AnCora corpus ([3]; [16]) and we extended the corresponding guidelines
([2]; [10]) in order to cover the specific phenomena of non-standard web texts.
In this way, we ensure the consistency and compatibility of the different Spanish
resources.

---

[2] http://nlp.lsi.upc.edu/freeling/

The main differences in the annotation scheme are due to the addition of special paratextual and paralinguistic tags for identifying and classifying the different types of phenomena occurring in this type of texts (misspellings, emphasis, repetitions, abbreviations, and punctuation transgressions, among others) and the criteria to be applied for dealing with them. However, the AnCora tagset has not been modified with new morphological or syntactic tags.

A summary of the criteria applied in the annotation of LAS-DisFo is presented below. We describe the criteria followed for word-level tokenization and its corresponding PoS tagging and then those applied for sentence-level tokenization and syntactic annotation.

5.1   Word-Level Tokenization

Most of the problems in the annotation process arose from word tokenization and word spellings. Therefore, the tokenization and morphological annotation processes required considerable effort. The kind of revision carried out consisted of addressing problems with word segmentation, verifying and assigning the correct PoS and lemma to each token, and resolving multiword expressions. The PoS annotation system[3] is based on [3].

Below, we present the criteria adopted in order to resolve the phenomena encountered in the discussion forum texts, which we have organized in the following groups: 1) word-segmentation phenomena; 2) typos and misspellings; 3) abbreviations; 4) marks of expressiveness; 5) foreign words, and 6) web items.

1. **Word-Segmentation Phenomena.** This kind of error mostly results from speed writing errors. As a general criterion, we always preserve the word form of the source text, except when the spelling error involves two different words with an incorrect segmentation, when the two words appear joined (1) or when a word is wrongly split due to the presence of a blank space (2). In these cases, the original text is modified. We justify this decision because this was a rare phenomenon, with an anecdotic presence in the corpus, and correcting these errors allowed for the correct PoS and syntactic annotation. In examples of word games,[4] we respect the original source and treat them like a multiword expression (if the words are split).

   (1) Esto **estan** de incrédulos. (instead of **es tan**)
   'This **isso** like incredulous people ...' (instead of **is so**)
   word=es lemma=ser pos=vaip3s
   word=tan lemma=tan pos=rg

   (2) Sistema de **gener ación** de bitcoins (instead of **generación**)
   'System for **gener ating** bitcoins' (instead of **generating**)
   word=generación lemma=generación pos=ncfs

---

[3] http://www.cs.upc.edu/~nlp/tools/parole-eng.html
[4] In the data annotated no word games were found.

In example (1) the criterion applied is to split the incorrect segment into two words, whereas in example (2) the criterion is to join the two segments into one word. In both cases, we assign the corresponding lemma and PoS tag to each word.

2. **Typos and Misspelling Errors.** The main typos found involve the omission/insertion of a letter, the transposition of two letters (3), the replacement of one letter for ano-ther, wrongly written capital letters, and proper nouns or any word that should be in capital letters but that appears in lower case. We also treat as typos those involving punctuation marks, usually a missing period in ellipsis (4).

(3) **pre**sonas '**pre**sons' (instead of **per**sona, '**per**son')
word=presonas lemma=persona pos=ncfp000 anomaly=yes

(4) pero.. lo bueno 'but.. the best thing' (instead of ...)
word=.. lemma= ... pos=fs anomaly=yes

In the case of misspellings, the most frequent mistakes are related to diacritic/accent removal, which normally also results in an incorrect PoS tag (5), but the omission of the silent 'h' in the initial position of the word, or the use of 'b' instead of 'v' (or vice versa), corresponding to the same phoneme, are also frequent. Dialectal variants (6), which are not accepted by the Royal Spanish Academy of Language, are also considered misspellings.

(5) todo cambio...'all change' (instead of todo cambi**ó** 'everything changed')
word=cambio lemma=cambiar pos=vmis3s anomaly=yes

(6) **a**moto (instead of moto, 'motorbike')
word=amoto lemma=moto pos=ncfs anomaly=yes

In example (5) the omission of the diacritic involves the assignment of an incorrect PoS, both 'cambio' and 'cambió' are possible words in Spanish, the former is a noun and the latter a verb, therefore the analyzer tagged 'cambió' as a noun. In this case, we manually assigned the correct verbal PoS (vmis3s) and the corresponding verbal lemma (infinitive, cambiar 'to change'), without modifying the original form.

The criteria adopted to resolve these phenomena is to maintain the source text, assign the correct PoS and lemma and add the label 'anomaly=yes' for typos and misspellings. In this way, the different written variants of the same word can be recovered through the lemma, and the typos and misspelling words are also easily identified by the corresponding labels.

3. **Abbreviations.** This kind of phenomena results in a simplification of the text aiming at reducing the writing effort. The abbreviations encountered

usually involve the omission of vowels, but consonants can also be omitted (7). In these cases, we assign the correct PoS and lemma and add the label 'toreviewcomment=a[5]' for identifying them.

(7) **tb** me gusta escribir 'I also like to write' (**tb** instead of también)
forma=tb lemma=también pos=rg toreviewcomment=a

4. **Marks of Expressiveness.** One of the phenomena that characterizes informal non-standard web texts is the unconventional use of graphic devices such as emoticons (8), capital letters (9) and (10), and the repetition of characters (11) to compensate for the lack of expressiveness in the writing mode. These are strategies that allow us to get closer to the direct interaction of oral communication. We use different labels and criteria to annotate the different types of marks of expressiveness:

For emoticons, we assign the lemma describing the emoticon with the prefix 'e-' and the PoS 'word', which indicates unknown elements.

(8) **:)**
word=:) lemma=e-contento ('e-happy') pos=word

For words in capital letters indicating emphasis (9) and for the emphatic repetition of vowels and other characters within words (10), we add the label 'polarity_modifier=increment'. We also assign the label 'toreviewcomment=cl[6]', when a fragment or an entire paragraph is written in capital letters (11). In this case, we add the label at the highest node (phrase or sentence).

(9) es algo totalmente **NUEVO!** 'is something totally **NEW!**'
word=NUEVO lemma=nuevo pos=aqms polarity_modifier=increment

(10) **muuuuy** grande**!!!** '**veeeery** big**!!!**' (instead of muy grande!)
word=muuuuy lemma=muy pos=rg polarity_modifier=increment
word=!!! lemma=! pos=fat polarity_modifier=increment

(11) LOS OTROS, LOS Q NO APORTAN, NO SE GANARÁN NI UN SEGUNDO D MI TIEMPO Y MI ESCRITURA.
'THE OTHERS, WHO DO NOT CONTRIBUTE ANYTHING, WILL NOT HAVE A SECOND OF MY TIME OR MY WRITING'.

(LOS OTROS, LOS Q NO APORTAN, NO SE GANARÁN NI UN SEGUNDO D MI TIEMPO Y MI ESCRITURA.) <sentence toreviewcomment=cl polarity_modifier=increment>

---

[5] 'a' stands for abbreviation.
[6] 'cl' stands for capital letters.

5. **Foreign Words.** In this kind of text the presence of words (12) or fragments written in another language (13), usually in English (and especially in technical jargon), is frequent. The criterion followed in these cases is not to translate the words to Spanish, and we add the label 'wdlng[7]=other'. In the case of fragments, we assign a simplified PoS tag (just the category) and all the words are grouped in a fragment at a top node (sentence, clause (S) or phrase).

(12) Estás **crazy**? 'Are you **loco**?' ('crazy' instead of loco)
word=crazy lemma=crazy pos=aqcs000 wdlng=other

(13) you are my brother
word=you lemma=you pos=p wdlng=other
word=are lemma=are pos=v wdlng=other
word=my lemma=my pos=t wdlng=other
word=brother lemma=brother pos=n wdlng=other

Syntactic annotation: (you are my brother)<sentence>

6. **Web Items**. We include website addresses, URLs, at-signs before usernames, and other special symbols used in web texts such as hashtags[8] in this category. Following the same criteria used in the AnCora annotation scheme, we tagged these web items as proper nouns and named entities with the value 'other'.

(14) http://www.afsca.gob.ar
word=http://www.afsca.gob.ar lemma= http://www.afsca.gob.ar pos=np
ne=other

5.2   Sentence Segmentation

The LAS-DisFo corpus was automatically sentence segmented in the PoS tagging process by the CLiC team, and the resulting segments were then manually corrected. It is worth noting that this level of segmentation required considerable human intervention because in informal web texts the use of punctuation marks frequently does not follow conventional rules: we found texts without any punctuation marks; texts that only used 'commas' as marks; texts with an overuse of strong punctuation marks usually for emphatic purposes, and texts with wrongly applied punctuation marks. These non-conventional uses lead to the erroneous automatic segmentation of sentences. Therefore, before starting with syntactic annotation it is necessary to correct the output of the automatic segmentation. The criteria followed are described hereafter.

---

[7] 'wdlng' stands for word language.
[8] In the data annotated no hashtags were found.

1. We apply normal sentence segmentation (15) when final punctuation (period, question mark, exclamation mark, or ellipsis) is correctly used. When ellipsis is used as non-final punctuation (16), we do not split the text.

   (15) (Hubieron dos detenidos por robos en medio del funeral...)<sentence> '(Two people were arrested for robberies in the middle of the funeral...')<sentence>

   (16) (Las necesidades no las crearon ellos solos... tambien ayudo el embargo)<sentence> '(The needs did not create themselves... it also helped the embargo')<sentence>

2. We do not split the text into separate sentences when final punctuation marks (usually periods) are wrongly used (17). If periods are used instead of colons, commas, or semicolons, we consider the text to be a sentence unit and we add the label 'anomaly=yes' to the punctuation mark.

   (17) (Los cambios que debería hacer Capitanich. Integrar publicidad privada. Cambiar a Araujo.)<sentence verbless=yes> '(The changes that Capitanich should make. Integrate private advertising. Switch to Araujo.)<sentence verbless=yes>'

   In example (17), the first period should be a colon and the second period should be a semicolon or a coordinated conjunction. In both cases, they are tokenized and tagged as periods (PoS=fp[9]) with the label 'anomaly=yes'. This sentence unit is treated as a <verbless> sentence because the main verb is missing.

   When the emoticons (18) are at the end of the sentence, they are included in the same sentence unit.

   (18) (Ni idea :?( )<sentence> '(No idea :?( )'<sentence>

3. We split the text into separate sentences when final punctuation marks are not included (19) and when a middle punctuation mark is used instead of final punctuation marks (20). In the former case, we add an elliptic node (∅) with the labels 'pos=fp', 'elliptic=yes' and 'anomaly=yes'. In the latter case, the label 'anomaly=yes' is added to the erroneous punctuation mark.

   (19) (Lo bueno debe prevalecer ∅<name=fp> <elliptic=yes> <anomaly=yes>) '(Good must prevail ∅<name=fp> <elliptic=yes> <anomaly=yes>))'

---

[9] 'fp' stands for punctuation period.

(20) hoy ya no pueden hacerlo, la tecnologia los mantiene a rayas,,
(hoy ya no pueden hacerlo, la tecnologia los mantiene a rayas, <PoS=fc>
<anomaly=yes> , <PoS=fc> <anomaly=yes>)<sentence>
'(today, they can no longer do so, the technology keeps them in line,
<PoS=fc> <anomaly=yes> , <PoS=fc> <anomaly=yes>)<sentence>

In example (20), the second comma could be interpreted either as an ellipsis
or as a repeated period. The context of this sentence points to the second
interpretation.

In addition to the commas incorrectly used as final punctuation marks, many
other problems appear in the sentence. In the example above the first word
of the sentence appears in lowercase instead of uppercase, the accent is missing in 'tecnologia' and 'rayas' should be written in singular (See section 5.1).

5.3 Syntactic Annotation

Regarding syntactic annotation, we followed the same criteria that we applied
to the AnCora corpus [16], following the basic assumptions described in [4]:
the annotation scheme used is theory-neutral; the surface word order is maintained and only elliptical subjects are recovered; we did not make any distinction
between arguments and adjuncts, so that the node containing the subject, that
containing the verb and those containing verb complements and adjuncts are
sister nodes.

We adopted a constituent annotation scheme because it is richer than dependency annotation (since it contains different descriptive levels) and, if it is necessary, it is easier to obtain the dependency structure from the constituent structure. Syntactic heads can be easily obtained from the constituent structure and
intermidiate levels can be avoided [5].

It was agreed to tag only those syntactic functions corresponding to sentence
structure constituents, whether finite or non-finite: only subject and verbal complements were taken into consideration. We defined a total number of 11 function tags, most of them corresponding to traditional syntactic functions: subject,
direct object, indirect object, prepositional object, adjunct, agent complement,
predicative complement, attribute, sentence adjunct, textual element and verbal
modifier.

When it was necessary to syntactically annotate more than one sentence
within a sentence unit (for instance, embedded clauses like relative, completive
and adverbial clauses), they were included under the top node <sentence>.
In the same way, embedded sentences were tagged as <S> with the feature
<clausetype> instantiated, its possible values being <completive>, <relative>,
<adverbial> and <participle>.

The syntactic annotation of LAS-DisFo did not present as large a variety
of phenomena as the morphological annotation did, but we did find many differences with respect to formal edited texts. In the discussion forum texts, the

Table 1. Syntactic information in LAS-DisFo and IS-NW

Corpus	Words	Sentences	Verbless	Discontinuities	Inserted elements
IS-NW	50,988	2,049	281	59	44
LAS-DisFo	50,291	2,846	1,229	139	98

frequency of verbless sentences, incomplete sentences, discontinuities and paranthetical elements (that did not belong to the general structure of the sentence) is higher than in news-based corpora such as IS-NewsWire[10]. Fragments without a main verb are treated as verbless sentences. In the case of LAS-DisFo, it is worth noting that these verbless sentences can be the result of joining several fragments of texts separated by wrongly used punctuation marks (17). Table 1 shows a comparision of these phenomena in the LAS-DisFo and LAS-NW corpora.

6   Final Remarks

In this paper, we have presented the criteria and annotation scheme followed in the morphological and syntactic annotation of the LAS-DisFo corpus, which contains 50,291 words and 2,846 sentences. Discussion Forum texts, like other kind of web texts, are characterized by an informal, non-standard style of writing. This results in texts with many misspellings and typographic errors and with a relaxation of the standard rules of writing. Furthermore, they usually contain pragmatic information about mood and feelings, often expressed by paratextual clues. All these characteristics pose difficult challenges to NLP tools and applications, which are designed for standard and formal written language.

The main challenges in the annotation of these kinds of texts appear in the segmentation of lexical and syntactic units and in the treatment of all the variants found at word level. To our knowledge, this is the first morphologically and syntactically annotated corpus of Spanish informal texts. This corpus will be released through the LDC catalog, and will be a new resource that could prove useful for deriving new tools for the analysis of informal Spanish language and Latin American Spanish, as well as for the linguistic analysis of spontaneous written language.

---

[10] The International Spanish Newswire TreeBank (IS-NW) consists of 50,988 words selected from the Spanish Gigaword previously released in LDC2011T12. The IS-NewsWire corpus has been also annotated with syntactic contituents and functions following the AnCora guidelines by the same annotator team. IS-NW and LAS-DisFo constitute the LDC Spanish Treebank, including the first Latin American Spanish corpus with morphological and syntactic annotation.

References

1. Bertran, M., Borrega, O., Martí, M.A., Taulé, M.: AnCoraPipe: A new tool for corpora annotation. Tech. rep., Working paper 1: TEXT-MESS 2.0 (Text-Knowledge 2.0) (2010). http://clic.ub.edu/files/AnCoraPipe_0.pdf
2. Bies, A., Mott, J., Warner, C., Kulick, S.: English Web Treebank. Linguistic Data Consortium, Philadelphia (2012)
3. Civit, M.: Criterios de etiquetación y desambiguación morfosintáctica de corpus en español. Coleccion de monografias de la SEPLN (2003)
4. Civit, M., Martí, M.A.: Design principles for a Spanish treebank. In: Proceedings of Treebanks and Linguistic Theories (2002)
5. Civit, M., Martí, M.A., Bufí, N.: Cat3LB and Cast3LB: from constituents to dependencies. In: Salakoski, T., Ginter, F., Pyysalo, S., Pahikkala, T. (eds.) FinTAL 2006. LNCS (LNAI), vol. 4139, pp. 141–152. Springer, Heidelberg (2006)
6. Dipper, S., Lüdeling, A., Reznicek, M.: NoSta-D: A corpus of German Non-Standard varieties. Non-standard DataSources in Corpusbased Research. Shaker Verlag (2013)
7. Foster, J.: "cba to check the spellig" investigating parser performance on discussion forum post. In: Proceedings of Human Language Technologies: The 2010 Annual Conference of the North American Chapter of the ACL, Los Angeles, California, pp. 381–384 (2010)
8. Foster, J., Çetinoglu, Ö., Wagner, J., Le Roux, J., Hogan, S., Nivre, J., Hogan, D., Van Genabith, J.: # hardtoparse: POS tagging and parsing the twitterverse. In: AAAI 2011 Workshop on Analyzing Microtext, pp. 20–25 (2011)
9. Garland, J., Strassel, S., Ismael, S., Song, Z., Lee, H.: Linguistic resources for genre-independent language technologies: user-generated content in BOLT. In: Proceedings of LREC 2012: 8th International Conference on Language Resources and Evaluation, Istanbul, Turkey (2012)
10. Maamouri, M., Bies, A., Kulick, S., Ciul, M., Habash, N., Eskander, R.: Developing an Egyptian Arabic treebank: impact of dialectal morphology on annotation and tool development. In: Proceedings of the Ninth International Conference on Language Resources and Evaluation (LREC 2014), Reykjavik, Iceland (2014)
11. Marcus, M., Kim, G., Marcinkiewicz, M., MacIntyre, R., Bies, A., Ferguson, M., Katz, K., Schasberger, B.: The penn treebank: annotating predicate argument structure. In: Proceedings of the Human Language Technology Workshop, San Francisco (1994)
12. Padró, L., Stanilovsky, E.: FreeLing 3.0: towards wider multilinguality. In: Proceedings of the Language Resources and Evaluation Conference (LREC 2012). ELRA, Istanbul, Turkey, May 2012
13. Petrov, S., McDonald, R.: Overview of the 2012 shared task on parsing the web. In: Notes of the First Workshop on Syntactic Analysis of Non-Canonical Language (SANCL), vol. 59. Citeseer (2012)
14. Seddah, D., Sagot, B., Candito, M., Mouilleron, V., Combet, V.: The French social media bank: a treebank of noisy user generated content. In: COLING 2012–24th International Conference on Computational Linguistics, Mumbai, pp. 2441–2458 (2012)

15. Song, Z., Bies, A., Riese, T., Mott, J., Wright, J., Kulick, S., Ryant, N., Strassel, S., Ma, X.: From light to rich ERE: annotation of entities, relations, and events. In: Proceedings of the 3rd Workshop on EVENTS: Definition, Detection, Coreference, and Representation. The 2015 Conference of the North American Chapter of the Association for Computational Linguistics - Human Language Technologies (NAACL HLT 2015), Denver (2015)
16. Soriano, B., Borrega, O., Taulé, M., Martí, M.A.: Guidelines: Constituents and syntactic functions. Tech. rep., Working paper: 3LB (2008). http://clic.ub.edu/corpus/webfm_send/17

Introduction of N-gram into a Run-Length Encoding Based ASCII Art Extraction Method

Tetsuya Suzuki$^{(\boxtimes)}$

Department of Electronic Information Systems,
Shibaura Institute of Technology, Saitama-shi, Saitama, Japan
tetsuya@shibaura-it.ac.jp

Abstract. As ASCII arts can be noise for natural language processing, ASCII art extraction methods can be used to remove them from text. A run-length encoding (RLE) based ASCII art extraction method proposed in our papers uses compression ratio by RLE for recognition of ASCII arts as ASCII arts tend to be compressed small by RLE and non-ASCII arts do not. It is because same characters tend to occur successively in ASCII arts but they do not in non-ASCII arts. Small ASCII arts, however, are not compressed as small as large ASCII arts. In this paper, we add the occurrence number of n-gram of ASCII arts in text into the RLE-based method as a new text attribute to cope with small ASCII arts. Our experimental results show that the new attribute improves the F-measure but it adds language-dependency into the RLE-based method though it is desirable that ASCII art extraction methods are language-independent.

Keywords: ASCII art · Pattern recognition · Natural language processing

1 Introduction

Text based pictures called *ASCII art* are often used in Web pages, email text and so on. ASCII arts are roughly classified into two major categories: the structure-based ASCII art and the tone-based ASCII art [10]. The structure-based ASCII art represents pictures where outlines of objects are drawn by characters. Fig. 1 shows a structure-based ASCII art of a smiley cat-like character. A simple smiley ':-)' is also categorized into the structure-based ASCII art. On the other hand, the tone-based ASCII art represents gray scale images consisting of characters. Fig. 2 is a tone-based ASCII art of a windmill. Recent ASCII arts use not only ASCII code characters but also Unicode characters, and have become more expressive.

ASCII arts, however, can be noise for natural language processing. For example, they are noises for text-to-speech software. Because a text-to-speech software can not ignore ASCII arts in text and pronounces digits and some symbols scattering in the ASCII arts, the speech confuses users.

ASCII art extraction methods, which detect areas of ASCII art in text, can solve the problem. ASCII art extraction methods can be constructed by *ASCII*

© Springer International Publishing Switzerland 2015
F. Daniel and O. Diaz (Eds.): ICWE 2015 Workshops, LNCS 9396, pp. 28–39, 2015.
DOI: 10.1007/978-3-319-24800-4_3

Fig. 1. A structure-based ASCII art   **Fig. 2.** A tone-based ASCII art

art recognition methods, which tell if a given fragment of text data is an ASCII art or not. With an ASCII art extraction method, we can ignore ASCII arts in text or replace them with other strings. It is desirable that ASCII art extraction methods are language-independent because a text may include one or more kinds of natural languages.

We proposed a run-length encoding (RLE) based ASCII art extraction method [4–7]. Because same characters tend to successively occur in lines of ASCII arts, compression ratio of text data by RLE represents how the text data looks like ASCII art. Small ASCII arts, however, are not compressed as small as large ASCII arts.

In this paper, we introduce n-gram of ASCII art into our ASCII art extraction method to cope with small ASCII arts, and evaluate effect of the introduction of n-gram by ASCII art extraction experiments.

The rest of the paper consists as follows. In section 2, we explain related work. In section 3 and 4, we explain the RLE-based method and introduce n-gram into the method respectively. In section 5, we explain our extraction experiments and show the results. In section 6, we evaluate the methods with n-gram by the experimental results. We finally state our conclusion in section 7.

2   Related Work

2.1   A Support Vector Machine-Based ASCII Art Recognition Method

Tanioka et al. proposed a support vector machine (SVM)-based ASCII art recognition method [1] as a natural language processing technique. Training data for SVM is a set of 262 dimension vectors. Each vector consists of two parts. The first 256 elements of the vector represent a *byte pattern*, whose i-th ($0 \leq i \leq 255$) element is the occurrence number of the byte data i in the byte stream of a UTF-8 encoded text. The authors categorized Japanese parts of speech into 6 groups. The rest 6 elements of the vector represent the occurrence numbers of the groups in text data. Because it is specific to Japanese language, this method will not work well for other languages.

2.2   A Byte Pattern Based Method

Nakazawa et al. proposed a byte pattern (BP) based method [2]. It scans a UTF-8 encoded text by one line and detects areas of ASCII art in the text. In advance of ASCII art extraction, an SVM model is constructed by a learning algorithm. Each training data consists of a class and an extended byte pattern. A class is either a class of ASCII art or a class of non-ASCII art. An extended byte pattern of a line is a concatenation of byte patterns of the line, the previous N-lines and the next N-lines. For example, if N is 1, an extended byte pattern of a line is a concatenation of byte patterns of three lines, which is a 768 dimension vector.

In ASCII art extraction, the ASCII art possibility of a line is calculated from an extended byte pattern of the line using the constructed SVM model. The BP-based method uses the smoothed ASCII art possibility to avoid splitting an ASCII art into parts. If the smoothed ASCII art possibility of a line is greater than or equal to 50%, the line is recognized as a part of ASCII art. The smoothed ASCII art possibility of a line is calculated from the ASCII art possibilities of the $2M+1$ lines, which are the line, the previous M-lines and the next M-lines.

According to our experimental comparison in [5], the RLE-based method and the BP-based method are competitive if training text and testing text are in a same set of languages, but the RLE based method works better than the BP-based method if training text and testing text are in different sets of languages.

3   A Run-Length Encoding Based Method

In this section, we explain our RLE-based method for ASCII art extraction [4, 6, 7]. We first explain two parts of the extraction method: a procedure called scanning with window width k and a procedure called text area reduction. We then explain the ASCII art extraction method. We finally explain an ASCII art recognition machine used in the ASCII art extraction method. We assume that texts are encoded in UTF-8.

3.1   Scanning with Window Width k

We define a procedure called *scanning with window width k*. Given a text data T, the procedure watches successive k lines on T and move the area from the beginning to the end of T by one line. We call the successive k lines as *a window*, and call the k as *the window width*. During the scanning, it applies a procedure, which is for text attribute extraction or ASCII art recognition, to each window.

3.2   Text Area Reduction

We define a procedure called *text area reduction* as follows. The procedure removes the following lines from a given text area represented by a pair of a start line and an end line of the entire text data.

- Successive lines from the start line which are recognized as non-ASCII art
- Successive lines from the end line which are recognized as non-ASCII art

This procedure uses an ASCII art recognition machine for the recognition.

3.3   An ASCII Art Extraction Method

We define an ASCII art extraction method with window width w. Given a text T, the procedure outputs a set of ASCII art areas in T with the ASCII art recognition machine M as follows.

1. It applies scanning with window width w to the text T with M. It means that the procedure applies M to the window in the scanning.
2. For each chunk of successive windows in which text data have been recognized as ASCII arts, it records the text data in the chunk of the windows as an ASCII art area candidate.
3. For each ASCII art area candidate, it applies the text area reduction procedure with M.
4. It outputs the results of the text area reduction procedure as a set of ASCII arts.

Fig. 3 shows an example of input text data for the RLE based extraction method, where an ASCII art is between English text. Fig. 4 shows an ASCII art area candidate obtained at the step 2 where there are redundant lines before and after the ASCII art. Fig. 5 shows the resulting ASCII art extracted at the step 3.

3.4   An ASCII Art Recognition Machine

We use an ASCII art recognition machine in the ASCII art extraction method, which is constructed by a machine learning algorithm. It takes a set of text attributes as its input and outputs whether true or false. The true value and the false value represent that the text is an ASCII art and that it is not respectively.

We construct training data for the machine learning as follows.

1. We prepare a set of ASCII arts and a set of non-ASCII arts.
2. We extract text attributes from them using the scanning with window width $k\ (= 1, 2, 3, \ldots, w)$.

The extracted text attributes are R, L and S. The attribute R is an attribute based on data compression ratio by RLE. Given a text T consisting of n lines, the attribute is defined as follow.

$$R \equiv \frac{\sum_{i=1}^{n} |RLE(l_i)|}{|T|} \qquad (1)$$

where $|x|$ denotes the length of a string x, $RLE(x)$ denotes a string encoded from the string x by RLE, and l_i is the i-th line of T. The attributes L and S are the number of lines and the length of the text data respectively.

Before we calculate the attributes of a given text, we normalize the text [4]. There exist two kinds of white spaces in Unicode, whose character codes are U+0020 and U+3000. We replace each U+3000 white space with two U+0020 white spaces in the text because the font width of U+3000 is the double of that of U+0020.

```
But perhaps we can run a scientific study of our own.
I'll volunteer for high IQ ('cause I'm an intellectual prick)
>>33 can volunteer for average IQ and
>>31 can volunteer for borderline-retarded IQ
Now let's go smoke dope._____
        /
｜ ひさしぶりだな
 ＼
     ＿＿｜／＿＿＿＿＿＿＿＿＿
     ∧_∧      ／
    （ ・∀・）   ∧∧ ＜  いいじゃないか
    （   ⊃）  （д゜：）  ＼＿＿＿＿＿＿＿＿＿＿＿＿
 ＿＿＿＿＿＿    （つ_つ___
     ̄ 日∇ ＼｜ BIBLO ｜＼
              ======    ＼
U.S. House of Representatives:
http://www.internationalrelations.house.gov/110/lee021507.htm
In the autumn of 1944, when I was 16 years old, my friend, Kim Punsun, and I were
collecting shellfish at the riverside when we noticed an elderly man and a Japanese man
looking down at us form the hillside......
A few days later, Punsun knocked on my window early in the morning,
and whispered to me to follow her quietly. I tip-toed out of the house after her.
```

Fig. 3. An input text for the RLE-based method

```
>>31 can volunteer for borderline-retarded IQ
Now let's go smoke dope._____
        /
｜ ひさしぶりだな
 ＼
     ＿＿｜／＿＿＿＿＿＿＿＿＿
     ∧_∧      ／
    （ ・∀・）   ∧∧ ＜  いいじゃないか
    （   ⊃）  （д゜：）  ＼＿＿＿＿＿＿＿＿＿＿＿＿
 ＿＿＿＿＿＿    （つ_つ___
     ̄ 日∇ ＼｜ BIBLO ｜＼
       ̄      ======    ＼
U.S. House of Representatives:
http://www.internationalrelations.house.gov/110/lee021507.htm
In the autumn of 1944, when I was 16 years old, my friend, Kim Punsun, and I were
```

Fig. 4. An ASCII art area candidate

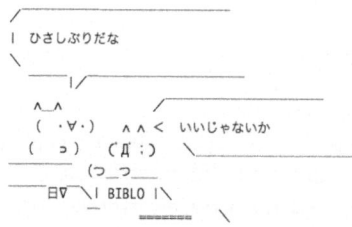

Fig. 5. A result of the RLE-based ASCII art extraction

4   Introduction of N-gram into the RLE-Based Method

The RLE-based method explained in section 3 assumes that same characters tend to successively occur in lines of ASCII arts but they do not in non-ASCII arts. Same characters, however, do not tend to successively occur in small ASCII arts.

To cope with such small ASCII arts, we add a text attribute N_g to the RLE-based method. Given a text T consisting of n lines, a set of n-grams N_{AA} in ASCII arts and a set of n-grams N_{nonAA} in texts without ASCII arts, the attribute N_g is defined as follows.

$$N_g \equiv |N_T - (N_{AA} - N_{nonAA})| \qquad (2)$$

where N_T is a set of n-grams in the text T.

5   Experiments

We compare the RLE-based method and the RLE-based method with n-gram by extraction tests. We implemented the two methods in Java. We used decision trees as ASCII art recognition machines for the two methods. The decision trees were constructed by the C4.5 machine learning algorithm [3] implemented in the data mining tool Weka [8,9] with the default parameters.

5.1   Texts

We used two sets of texts E and J encoded in UTF-8 for the machine learning. The set of texts E consists of English texts with 289 ASCII arts and 290 non-ASCII arts, whose lines range from 1 to 118. The set of texts J consists of Japanese texts with 259 ASCII arts and 299 non-ASCII arts, whose lines range from 1 to 39. Their new line code is the sequence of CR LF.

We constructed training data and testing data as follows. We divided the set of texts E and J into two groups A and B. Each of A and B consists of English texts and Japanese texts. We then made 800 texts from A. Each of the 800 texts consists of three parts X, Y and Z where X and Z are randomly selected non-ASCII art from A and Y is randomly selected ASCII art from A. Each of X, Y and Z is English or Japanese text data. There exist eight combinations of languages for X-Y-Z: (1) E-E-E, (2) E-E-J, (3) E-J-E, (4) E-J-J, (5) J-E-E, (6) J-E-J, (7) J-J-E and (8) J-J-J where E and J means English and Japanese respectively. Figure 3 shows an example of such testing data where X and Z are English texts and Y is a Japanese ASCII art. For each of the eight combinations, we made 100 texts. As a result, we got 800 texts in total. We also made 800 texts from B in the same way.

5.2 Conditions

ASCII art extraction experiments by the two methods were conducted under the following conditions.

We measured the average of precision, the average of recall and the average of F-measure in ASCII art extraction by 2-fold cross validation using the two sets A and B in the following five cases.

E-J Training text data is in English, and testing text data is in Japanese.
J-E Training text data is in Japanese, and testing text data is in English.
E-E Training text data is in English, and testing text data is also in English.
J-J Training text data is in Japanese, and testing text data is also in Japanese.
EJ-EJ Training text data is in English and Japanese, and testing text data is also in English and Japanese.

We changed the window width from 1 to 10 in ASCII art extraction.

In the RLE-based method with n-gram, we used all bigrams extracted from the two group A and B as n-grams. We extracted 2714 bigrams in English text and 5275 bigrams in Japanese text from the group A, and extracted 2155 bigrams in English text and 5921 bigrams in Japanese text from the group B.

5.3 Results

The experimental results are shown in Table 1, 2, 3, 4 and 5. In the tables, the results by the RLE-based method are shown in the columns titled "R, L, S", and the results by the RLE-based method with the attribute N_g are shown in the columns titled "R, L, S, N_g". The two titles are sets of attributes which the two methods used.

Table 1 shows the results under the case E-J. In the case of the RLE-based method, the highest average of F is 0.897 when the widow width is 2. In the case of the RLE-based method with n-gram, the highest average of F is 0.901 when the widow width is 4. The highest average of F in the RLE-based method with n-gram is higher by 0.004 points compared to that in the RLE-based method.

Table 2 shows the results under the case J-E. In the case of the RLE based method, the highest average of F is 0.983 when the widow width is 8 and 9. In the case of the RLE based method with n-gram, the highest average of F is 0.968 when the widow width is 8. The highest average of F in the RLE-based method with n-gram is lower by 0.015 points compared to that in the RLE-based method.

Table 3 shows the results under the case E-E. In the case of the RLE-based method, the highest average of F is 0.984 when the widow width is 4 and 5. In the case of the RLE-based method with n-gram, the highest average of F is 0.993 when the widow width is 5. The highest average of F in the RLE-based method with n-gram is higher by 0.009 points compared to that in the RLE-based method.

Table 4 shows the results under the case J-J. In the case of the RLE-based method, the highest average of F is 0.926 when the widow width is 4. In the case

Table 1. The results in the case E-J

Window Width	R, L, S			R, L, S, N_g		
	Avg. of Precision	Avg. of Recall	Avg. of F	Avg. of Precision	Avg. of Recall	Avg. of F
1	0.897	0.762	0.819	0.887	0.778	0.818
2	0.907	0.892	0.897	0.919	0.894	0.896
3	0.868	0.887	0.880	0.889	0.908	0.886
4	0.851	0.892	0.873	0.902	0.911	0.901
5	0.858	0.902	0.879	0.839	0.910	0.858
6	0.790	0.904	0.838	0.817	0.924	0.856
7	0.774	0.916	0.829	0.853	0.912	0.875
8	0.771	0.909	0.825	0.850	0.923	0.876
9	0.740	0.920	0.805	0.854	0.909	0.887
10	0.736	0.913	0.801	0.844	0.909	0.880

Table 2. The results in the case J-E

Window Width	R, L, S			R, L, S, N_g		
	Avg. of Precision	Avg. of Recall	Avg. of F	Avg. of Precision	Avg. of Recall	Avg. of F
1	0.981	0.880	0.943	0.944	0.817	0.914
2	0.984	0.934	0.969	0.936	0.841	0.913
3	0.985	0.946	0.976	0.910	0.837	0.895
4	0.984	0.954	0.980	0.966	0.945	0.958
5	0.985	0.958	0.982	0.962	0.950	0.964
6	0.979	0.971	0.982	0.960	0.952	0.961
7	0.975	0.973	0.980	0.941	0.938	0.952
8	0.975	0.973	0.983	0.957	0.966	0.968
9	0.974	0.975	0.983	0.956	0.967	0.966
10	0.971	0.970	0.981	0.951	0.958	0.964

of the RLE-based method with n-gram, the highest average of F is 0.982 when the widow width is 4. The highest average of F in the RLE-based method with n-gram is higher by 0.056 points compared to that in the RLE-based method.

Table 5 shows the results under the case EJ-EJ. In the case of the RLE-based method, the highest average of F is 0.957 when the widow width is 6. In the case of the RLE-based method with n-gram, the highest average of F is 0.984 when the widow width is 4, 5 and 7. The highest average of F in the RLE-based method with n-gram is higher by 0.027 points compared to that in the RLE-based method.

Table 3. The results in the case E-E

Window Width	R, L, S			R, L, S, N_g		
	Avg. of Precision	Avg. of Recall	Avg. of F	Avg. of Precision	Avg. of Recall	Avg. of F
1	0.992	0.897	0.956	0.997	0.931	0.962
2	0.993	0.945	0.983	0.994	0.963	0.983
3	0.983	0.951	0.981	0.991	0.985	0.988
4	<u>0.985</u>	<u>0.956</u>	<u>0.984</u>	0.993	0.991	0.992
5	<u>0.983</u>	<u>0.957</u>	<u>0.984</u>	<u>0.992</u>	<u>0.986</u>	<u>0.993</u>
6	0.977	0.960	0.982	0.988	0.984	0.990
7	0.971	0.963	0.980	0.986	0.995	0.989
8	0.968	0.963	0.979	0.986	0.997	0.990
9	0.965	0.963	0.977	0.984	0.997	0.989
10	0.965	0.964	0.977	0.981	0.997	0.987

Table 4. The results in the case J-J

Window Width	R, L, S			R, L, S, N_g		
	Avg. of Precision	Avg. of Recall	Avg. of F	Avg. of Precision	Avg. of Recall	Avg. of F
1	0.918	0.771	0.834	0.957	0.830	0.882
2	0.928	0.879	0.904	0.976	0.946	0.957
3	0.938	0.886	0.919	0.980	0.972	0.975
4	<u>0.941</u>	<u>0.901</u>	<u>0.926</u>	<u>0.984</u>	<u>0.974</u>	<u>0.982</u>
5	0.927	0.888	0.916	0.979	0.967	0.978
6	0.893	0.871	0.899	0.973	0.966	0.977
7	0.894	0.877	0.909	0.970	0.967	0.974
8	0.886	0.878	0.905	0.966	0.954	0.971
9	0.868	0.866	0.891	0.960	0.948	0.971
10	0.868	0.862	0.890	0.948	0.937	0.961

Table 5. The results in the case EJ-EJ

Window Width	R, L, S			R, L, S, N_g		
	Avg. of Precision	Avg. of Recall	Avg. of F	Avg. of Precision	Avg. of Recall	Avg. of F
1	0.953	0.846	0.893	0.974	0.877	0.919
2	0.966	0.915	0.946	0.984	0.942	0.962
3	0.966	0.919	0.949	0.989	0.974	0.980
4	0.959	0.928	0.953	<u>0.988</u>	<u>0.979</u>	<u>0.984</u>
5	0.957	0.929	0.950	<u>0.984</u>	<u>0.976</u>	<u>0.984</u>
6	<u>0.954</u>	<u>0.928</u>	<u>0.957</u>	0.981	0.976	0.983
7	0.951	0.922	0.950	<u>0.982</u>	<u>0.974</u>	<u>0.984</u>
8	0.944	0.920	0.951	0.978	0.975	0.983
9	0.935	0.920	0.945	0.974	0.973	0.980
10	0.920	0.919	0.936	0.972	0.975	0.980

Table 6 shows the number of ASCII arts which the two ASCII art extraction methods can not extract well from testing data with the best window widths in the case EJ-EJ regardless of non-ASCII arts before and after the ASCII arts. There are 22 kinds of ASCII arts in testing data with F-measures in the range from 0.0 to 0.1 but not in testing data with F-measures in the range from 0.9 to 1.0 when the RLE-based method extracts ASCII arts with the window width 6. The number of lines of them ranges from 1 to 51. On the other hand, there are 4 kinds of ASCII arts in testing data with F-measures in the range from 0.0 to 0.1 but not in testing data with F-measures in the range from 0.9 to 1.0 when the RLE-based method with n-gram extracts ASCII arts with the window width 4. The number of lines of them ranges from 2 to 3. The 4 kinds of ASCII arts, which are small structure-based ASCII arts, are included in the 22 kinds of ASCII arts. One of the 4 ASCII arts is shown in Fig. 6.

Table 6. The number of ASCII arts difficult to be extracted in the cases EJ-EJ

The number of lines of ASCII arts	R, L, S	R, L, S, N_g
1	3	0
2	3	2
3	3	2
4	4	0
5	2	0
8	2	0
11	2	0
12	1	0
28	1	0
51	1	0

```
(・ω・)  ナデナデして～
(nn)
```

Fig. 6. One of the ASCII arts difficult to be extracted

6   Evaluation

We can summarize the experimental results shown in section 5.3 as follows. The three cases of E-E, J-J and EJ-EJ are cases where training texts and testing texts are in a same set of languages. In these cases, the highest average of F-measures in the RLE-based method with n-gram is higher by 0.009 points to 0.056 points compared to that in the RLE-based method. The two cases of E-J and J-E are cases where training texts and testing texts are in different sets of languages. In the case of E-J, the highest average of F-measures in the RLE-based method with n-gram is higher by 0.004 points compared to that in the RLE-based method. In the case of J-E, however, the highest average of F-measures in the RLE-based method with n-gram is lower by 0.015 points compared to that in the RLE-based method. The new attribute based on n-grams contributes to extracting not only small ASCII arts but also large ASCII arts.

The reason why the RLE-based method does not use any word dictionary is to make the method language-independent. It is desirable that ASCII art extraction methods are language-independent because a text may include one or more kinds of natural languages.

The introduction of n-gram into the RLE-based method, however, adds language-dependency into the method though it improves the F-measure in the RLE-based method in the case when training texts and testing texts are in a same set of languages.

7   Conclusion

We introduced the occurrence number of n-grams of ASCII art into a run-length encoding (RLE) based ASCII art extraction method as a new text attribute to cope with small ASCII arts. It is because compression ratio of texts by RLE works well to distinguish if a text is an ASCII art or not for large ASCII arts but it does not work well for small ASCII arts. The RLE-based method uses a machine learning algorithm for pattern recognition of ASCII art. Our experimental results show that the introduction of the new text attribute improves the F-measure in the RLE-based method in the case when training texts and testing texts are in a same set of languages. It, however, adds language-dependency into the RLE-based method though it is desirable that ASCII art extraction methods are language-independent because a text may include one or more kinds of natural languages.

References

1. Hiroki, T., Minoru, M.: Ascii Art Pattern Recognition using SVM based on Morphological Analysis. Technical report of IEICE. PRMU 104(670), 25–30 (20050218). http://ci.nii.ac.jp/naid/110003275719/
2. Nakazawa, M., Matsumoto, K., Yanagihara, T., Ikeda, K., Takishima, Y., Hoashi, K.: Proposal and its evaluation of ASCII-art extraction. In: Proceedings of the 2nd Forum on Data Engineering and Information Management (DEIM2010), pp. C9–C4 (2010)
3. Quinlan, J.R.: C4.5: Programs for Machine Learning. Morgan Kaufmann Publishers Inc., San Francisco (1993)
4. Suzuki, T.: A comparison of whitespace normalization methods in a text art extraction method with run length encoding. In: Deng, H., Miao, D., Lei, J., Wang, F.L. (eds.) AICI 2011, Part III. LNCS, vol. 7004, pp. 135–142. Springer, Heidelberg (2011). http://dx.doi.org/10.1007/978-3-642-23896-3_16

5. Suzuki, T.: Comparison of two ASCII art extraction methods: a run-length encoding based method and a byte pattern based method. In: Proceedings of the 6th IASTED International Conference on Computational Intelligence. ACTA Press (2015)
6. Suzuki, T., Hayashi, K.: A language-independent text art extraction method. In: Proceedings of the 2nd International Conference on the Applications of Digital Information and Web Technologies, pp. 462–467. IEEE Computer Society (2009)
7. Suzuki, T., Hayashi, K.: Text data compression ratio as a text attribute for a language-independent text art extraction method. In: Proceedings of the 3rd International Conference on the Applications of Digital Information and Web Technologies (2010)
8. The University of Waikato: Weka 3 - Data Mining with Open Source Machine Learning Software in Java. http://www.cs.waikato.ac.nz/ml/weka/ (retrieved on December 14, 2008)
9. Witten, I.H., Frank, E.: Data Mining: Practical Machine Learning Tools and Techniques, 2nd edn. Morgan Kaufmann (2005)
10. Xu, X., Zhang, L., Wong, T.T.: Structure-based ASCII Art. ACM Transactions on Graphics (SIGGRAPH 2010 issue) **29**(4), 52:1–52:9 (2010)

SMT: A Case Study of Kazakh-English Word Alignment

Amandyk Kartbayev[(✉)]

Laboratory of Intelligent Information Systems,
Al-Farabi Kazakh National University, Almaty, Kazakhstan
a.kartbayev@gmail.com

Abstract. In this paper, we present results from a set of experiments to determine the effect on translation quality, it depends on the particular kind of morphological preprocessing that can be represented by finite-state transducers. A high agglutinative nature of the Kazakh language under the condition of poor language resources makes an issue in the processing of derivational morphology. Our methods are focused on useful phrase pairs in word alignment and provide a most language independent approach, which may improve a translation into other morphological complex languages. We processed our algorithms over the Kazakh Wikipedia base of about 1.5 million unique lexeme and 230 million words overall. Our best translation system increases 3 BLEU points over the Kazakh-English baseline on a blind test set.

Keywords: Word alignment · Kazakh morphology · Word segmentation · Machine translation

1 Introduction

In this work we focus on the word alignment process, which is the most important part of information recovery from a source with a lot of inflection. Particularly, we are interested in the sources where the given sentence pairs contain more new words with a less prior information about their nature. This is a challenging problem in machine translation and it is a hard to learn from the lexicon and usually repeats the similar errors again and again.

Morphological segmentation process intended to break words into morphemes, which are the basic semantic units and a key component for natural language processing systems. This is our current subtask in the machine translation project and we also desired to show that a simple segmentation scheme can perform pretty well as the most sophisticated one.

Most papers in statistical machine translation (SMT) oriented morphology analysis presents experiments that they consist of numerous experimentation to choose the best among a set of segmentation schemes. These morphological preprocessing schemes focused on various level of decomposition and compare the resulting translation performances, but usually use a subset of morphology and apply only a few simple rules in a segmentation process.

© Springer International Publishing Switzerland 2015
F. Daniel and O. Diaz (Eds.): ICWE 2015 Workshops, LNCS 9396, pp. 40–49, 2015.
DOI: 10.1007/978-3-319-24800-4_4

In the paper, well known to the intended audience, El-Kahlout and Oflazer[1] explored this task for English to Turkish translation, which is an agglutinative language as Kazakh. Their methods used in the survey were a morphological analyzer and token disambiguation, though translation models trained throw morphemes obviously degrades the translation quality. But they outperformed the baseline results after some morpheme grouping techniques. A research more relevant to this work was done by Bisazza and Federico[2].

Our segmentation model incorporates simple ideas inspired by finite state features such as morphemes and their contexts in the range of situations, where the lexeme is likely a morpheme, as any other cases, it is a word boundary. We develop a segmentation scheme using syntactic and morphological rules are implemented as finite-state transducers. We focus on derivational morphology and tested our approach on Kazakh wiki and news datasets, which was crawled from Web. The affix system, which will be the focus in this paper, is described in more detail in Table 1.

Table 1. An example of Kazakh agglutination

Stem	Plural affixes	Possesive affixes	Case affixes
stem[kol'+]	plural[+der]	1-st pl.[+imiz]	locative[+de]
stem[kol'+]	-	1-st s.[+im]	locative[+de]
stem[kol'+]	-	-	locative[+de]

Our system, using monolingual features only, is one of the most realistic application for Kazakh and compared to Morfessor tool[3], so it can be readily applied to supervised and semi-supervised learning of morphological inflations of the language even on speech processing. Also morphological adjustment gives a improved statistical machine translation performance over the pair of the morphological rich and poor languages. A substantial improvement in translation performance is achieved, when we used word alignments learned from the output of the processing technique, but we found that some of the segmentation errors are caused by morphological analyzer. These kind of errors could be avoided using data selection, which demonstrates the ability of the method fix it successfully. Using morphological analysis we out grammatical features of word and can find syntactic structure of input sentence, which further demonstrates the benefit of using this method in machine translation.

In this paper, we present a systematic comparison of preprocessing techniques for a Kazakh-English pair. Previous researches that we explored on our approaches are rule-based morphological analyzers[4], which consist in deep language expertise and a exhaustive process in system development. Unsupervised approaches use actually unlimited supplies of text to cover very few labeled resource and it has been widely studied for a number of languages[5]. However, existing systems are too complicated to extend them with random overlapping dependencies that are crucial to segmentation.

On our general task we refer to the methodology exposed by Oflazer and El-Kahlout on the Turkish-English task. Because, Turkish is also morphological rich language like Kazakh and not all affix combinations looks grammatical. This means the linguistic knowledge is the key to finding significant segmentation schemes among many possible combinations of the rules. Only rule-based approaches are provided and have done detailed analyses of the Kazakh morphological parsing task. For a comprehensive survey of the rule-based morphological analyze we refer a reader to the research by Altenbek[6] and Kairakbay[7].

The paper is structured as follows: Section 2 discusses the key challenges of translating Kazakh to English. In Section 3 we described the different segmentation techniques we study. And Section 4 presents our evaluation results.

2   Translation Task

In our work, we experiment with a range of segmentation technique totally giving a five best distinct schemes. Our results show that the proper selection of the segmentation scheme has a significant impact on the performance of a phrase-based system in a large corpora. The translation experiments described in this paper are carried out with a standard phrase-based Moses[8] system (not with Experiment Management System) and the target-side language models were trained on the MultiUN[9] corpora.

Generally, breaking up a process of generating the data into smaller steps, modeling the smaller steps with probability distributions, and combining the steps into a coherent story is called generative modeling. The phrase-based models are generative models that translate sequences of words in f_j into sequences of words in $e_j(1)$, in difference from the word-based models that translate single words in isolation.

$$P\left(e_j \mid f_j\right) = \sum_{j=1}^{J} P\left(e_j, a_j \mid f_j\right) \tag{1}$$

Improving translation performance directly would require training the system and decoding each segmentation hypothesis, which is computationally impracticable. That we made various kind of conditional assumptions using a generative model and decomposed the posterior probability(2). In this notation e_j and f_i point out the two parts of a parallel corpus and a_j marked as the alignment hypothesized for f_i.

$$P\left(e_j^J, a_j^J \mid f_i^I\right) = \frac{f_i}{(I+1)^J} \prod_{j=1}^{J} p\left(e_j \mid f_{a_j}\right) \tag{2}$$

The use of phrases as translation units is motivated by the observation that sometimes one word in a source language translates into multiple words.

Because, a Kazakh word can correspond to a single English word, up to phrases of various lengths, or even to a whole sentence as shown in Table 2.

Our objective is to produce alignments, which can be used to build high quality machine translation systems[10]. These are pretty close to human annotated alignments that often contain m-to-n alignments, where several source words are aligned to several target words and the resulting unit can not be further decomposed. Using segmentation, we describe a new generative model which directly models m-to-n non-consecutive word alignments. There is a very small improvement in alignment if a source word occurs only once in the parallel text, the probability assigned to it, generates each of the words to the each sentence, will be too high. This problem is solved by smoothing the word-to-word translation probabilities with a coincident distribution.

3   Improving Word Alignment

In order to look through this task, we did a series of experiments and found morpheme alignment can be employed to increase the similarity between languages, therefore enhancing the quality of machine translation for Kazakh-English language pair. Our experiments consist of two parts: one is on Kazakh-English morphological segmentation; the other is a case study of the benefits of morpheme based alignment.

We use following heuristic methods that improve the generative models for phrase alignment. At first, the tags were assigned to the obtained phrase pair pieces, then we make classification and clustering the phrases according their contexts, also we extract phrase pairs that are not linked within the word alignments, like the phrases containing multiword entities that can not be correctly aligned. We obtained word alignments in both translation directions by the GIZA++ toolkit[11], which is based on the IBM models[12]. We prefer a grow-diag-final symmetrization method to others for both alignment directions.

As the first part of our experiments we morphologically segmented Kazakh input sentences to compute morpheme alignment. For these purposes we used Morfessor, an unsupervised analyzer and Helsinki Finite-State Toolkit (HFST)[13]. Helsinki Finite State Toolkit is an open-source implementation of the Xerox finite-state toolkit, that implements the lexc, twol and xfst formalisms for modeling morphlogical rules. After these Kazakh stems and suffixes is converted into labeled morphemes, as well as particular English verbs. We append a plus sign to the end of each open tag to know boundaries of internal morphemes from final ones, e.g., [stem+] and [stem] are assumed as different tokens.

3.1   Morphological Segmentation

Our preprocessing job starts from morphological segmentation, which includes running Morfessor tool and HFST to each entry of the corpus dictionary. The first step of word segmentation aims to get suffixes and roots from a vocabulary consisting of 1500k unique word forms taken from Kazakh Wikipedia dump[14].

Accordingly, we take surface forms of the words and generate their all possible lexical forms.

In the Kazakh language, as in other agglutinative languages,the morphemes are affixed to the root due to the morphotactic rules of the language. These morphotactic rules define the states and the suffixes that can be added to the stem, then change the state of the affixed word. These rules often represented by the certain finite state transducers. Where the transitions are marked as the derivational morphemes, that come in same order as the affixation of the word. Also we use the lexicon to label the initial states as the root words by parts of speech such as noun, verb, etc. The final states represent a lexeme created by affixing morphemes in each further states.

The schemes presented below are different combinations of outputs determining the removal of affixes from the analyzed words. The baseline approach is not perfect since a scheme includes several suffixes incorrectly segmented. In this case, we mainly focused on detection a few techniques for the segmentation of such word forms. In order to find an effective rule set we tested several segmentation schemes named S[1..5], some of which have described in the following Table 2.

Table 2. The segmentation schemes

Id	Schema	Examples	Translation
S1	stem	el	state
S2	stem+case	el + ge	state + dative
S3	stem+num+case	el + der + den	state + num + ablativ
S4	stem+poss+	el + in	state + poss2sing
S5	stem+poss+case	el + i +ne	state + poss3sing + dative

Nominal cases that are expected to have an English counterpart are split off from words: these are namely dative, ablative, locative and instrumental, often aligning with the English prepositions to, from, in and with/by. The remaining case affixes nominative, accusative and genitive are not have English counterparts. After treating case affixes we split of possessive suffixes from nouns of all persons except the 1st singular, which doesnt need removed.

There are large amount of verbs presenting ambiguity during segmentation, as suppositional verbs 'eken' - 'to seem' and 'goi'. Which do not take personal endings, but follow conjugated main verbs. The verb 'to become' has the forms 'bolu' - 'to become', 'bolar' - 'will become', and 'bolmau' - 'to not become'. There are also the verbs 'bar' - 'to exist/have' and 'jok' - 'to not exist/not have'. These are special verbs because they do not take personal endings. Also a verbs generally refer to group action, e.g. 'oinasu' - 'to play together', 'soilesu' - 'to converse' produce an ambiguity, e.g. a stem 'soile' - 'say' and a suffix 'su' - 'water'. During the process, we hardly determined the border between stems and inflectional affixes, especially when the word and the suffix matches entire word in the language. For instance, a progressive auxiliary word 'jat' - 'alien' and the negation morphemes like 'ba', 'ma', etc, though an irregular form of several verbs. Under many situations, the type of words, which we described, made an inaccurate stemming. In fact, there are lack of syntactic information we cannot easily distinguish among similar cases.

While GIZA++ tool produces a competitive alignment between words, the Kazakh sentences must be segmented as we already have in the first step. Therefore our method looks like an word sequence labeling problem, the contexts can be presented as POS tags for the word pairs.

Table 3. Part of Speech tag patterns

Tag	Sample	Tag	Sample
NN (Noun)	"el"-"state"	JJS (Adjective, super.)	"tym"-"most"
NNP (Proper noun)	"biz"-"we"	VB (Verb, base form)	"bar"-"go"
JJ (Adjective)	"jasyl"-"green"	VBD (Verb, past tense)	"bardy"-"went"
JJR (Adj, comp.)	"ulkenirek"-"bigger"	VBG (Verb, gerund)	"baru"-"to go"
RB (Adverb)	"jildam"-"speedy"	CC (Conjunction)	"jane"-"and"

3.2   Alignment Model

We extend the alignment modeling process of Brown et al. at the following way. We assume the alignment of the target sentence e to the source sentence f is a. Let c be the tag(from Penn Treebank) of f for segmented morphemes. This tag is an information about the word and represents lexeme after a segmentation process. This assumption is used to link the multiple tag sequences as hidden processes, that a tagger generates a context sequence c_j for a word sequence f_j(3).

$$P\left(e_1^I, a_1^I \mid f_1^J\right) = P\left(e_1^I, a_1^I \mid c_1^J, f_1^J\right) \tag{3}$$

Then we can show Model 1 as(4):

$$P\left(e_i^I, a_i^I \mid f_j^J, c_j^J\right) = \frac{1}{(J+1)^I} \prod_{i=1}^{I} p\left(e_i \mid f_{a_i}, c_{a_i}\right) \tag{4}$$

The training is carried out in the tagged Kazakh side and the untagged English side of the parallel text. If we estimate translation probabilities for every possible context of a source word, it will lead to problems with data sparsity and rapid growth of the translation table. We applied expectation maximization(EM) algorithm to cluster a context of the source sentence using similar probability distributions, avoiding problems with data sparsity and a size of the translation table another case.

We estimate the phrase pairs that are consistent with the word alignments, and then assign probabilities to the obtained phrase pairs. Context information is incorporated by the use of part-of-speech tags in both languages of the parallel text, and the EM algorithm is used to improve estimation of word-to-word translation probabilities. The probability p_k of the word w to the corresponding context k is:

$$p_k(w) = \frac{p_k f_k(w \mid \phi_k)}{\sum p_i f_i(w \mid \phi_i)} \tag{5}$$

Where, ϕ is the covariance matrix, and f are certain component density functions, which evaluated at each cluster. After we use association measures to filter infrequently occurring phrase pairs by log likelihood ratio r estimation[15]. For n pairs of the phrases, we can obtain the phrase pairs whose comparative values are larger than a threshold value as follows(6):

$$R(f,e) = \frac{r(f,e)}{Max_e r(f,e)} \tag{6}$$

Our algorithm, like a middle tier component, processes the input alignment files in a single pass. Current implementation reuses the code from https://github.com/akartbayev/clir that conducts the extraction of phrase pairs and filters out low frequency items. After the processing all valid phrases will be stored in the phrase table and be passed further. This algorithm proposes refinement by adding morphological constraints between the direct and the reverse directions of the alignment, which may improve the final word alignments.

4   Evaluation

Though our final objective is an improvement of the translation quality of SMT systems, we evaluate the alignment relies with the phrase-based system on the Kazakh-English parallel corpus of approximately 60K sentences, which have a maximum of 100 morphemes. Our corpora consists of the legal documents from http://adilet.zan.kz, a content of http://akorda.kz, and Multilingual Bible texts. We conduct all experiments on a single PC, which runs the 64-bit version of Ubuntu 14.10 server edition on a 4Core Intel i7 processor with 32 GB of RAM in total. All experiment files were processed on a locally mounted hard disk. Also we expect the more significant benefits from a larger training corpora, therefore we are in the process of its construction.

We did not have a gold standard for phrase alignments, so we had to refine the obtained phrase alignments to word alignments in order to compare them with our word alignment techniques.

Table 4 shows the change in alignment error rate (AER) of the alignments, that the improved model produce a decrease in AER and leads to a better translation quality, measured by BLEU score[16]. A high recall apparently improves translation quality, but low precision may decrease it and a relation between recall and precision is substantial. A high recall and low precision in alignment

Table 4. Alignment quality results

System	Precision	Recall	F-score	AER
Baseline	57.18	28.35	38.32	36.22
Morfessor	71.12	28.31	42.49	20.19
Rule-based	89.62	29.64	45.58	09.17

Table 5. Metric scores for all systems

System	BLEU	METEOR	TER
Baseline	30.47	47.01	49.88
Morfessor	31.90	47.34	49.37
Rule-based	33.89	49.22	48.04

pretty significant for the amount of generated phrases. The best situation takes place on well maintained recall and precision, which is a result of our study.

We employed an approach of the morpheme-based representation as explained in Section 3 about the morphological analysis, which impacts an improvement of +2 BLEU points. The system parameters were optimized with the minimum error rate training (MERT) algorithm [17], and evaluated on the out-of and in-domain test sets. Monolingual corpora from News Commentary was partially used, when we trained 5-gram language models. All language models were trained with the IRSTLM toolkit[18] and then were converted to binary form using KenLM for a faster execution[19].

Table 5 visualizes the best BLEU scores, which were computed using the MultEval[20]: BLEU, TER[21] and METEOR[22]; and we ran Moses three times per experiment setting, and report the highest BLEU scores obtained. Our survey shows that translation quality measured by BLEU metrics is not strictly related with lower AER.

5   Conclusions

In this work, we address a morpheme alignment problems concerned highly inflected languages. We compared our approach against a baseline of the Moses translation pipeline and another common approach to inflected languages segmentation. By using our method for phrase selection we were able to obtain translation quality better than the baseline method produce, while the phrase table size and the noise phrase pairs have been reduced by substantial level. Although memory requirements of the processing environment are increased, but they are still within manageable limits.

Our method is comparable to other language-specific works, and there are many possible directions for future research. As our approach may produce improvements in alignment quality, any downstream changes of the translation model also possible. We learned that processing the features are integrated into the standard phrase table is an area for improvement. That was our initial investigation into alignment models and further translation experiments will be carried out.

References

1. Oflazer, K., El-Kahlout, D.: Exploring different representational units in English-to-Turkish statistical machine translation. In: 2nd Workshop on Statistical Machine Translation, Prague, pp. 25–32 (2007)
2. Bisazza, A., Federico, M.: Morphological pre-processing for Turkish to English statistical machine translation. In: International Workshop on Spoken Language Translation 2009, Tokyo, pp. 129–135 (2009)
3. Creutz, M., Lagus, K.: Unsupervised models for morpheme segmentation and morphology learning. ACM Transactions on Speech and Language Processing **4**, article 3. Association for Computing Machinery, New York (2007)
4. Beesley, K.R., Karttunen, L.: Finite State Morphology. CSLI Publications, Palo Alto (2003)
5. Goldsmith, J.: Unsupervised learning of the morphology of a natural language. Computational Linguistics **27**, 153–198 (2001)
6. Altenbek, G., Xiao-Long, W.: Kazakh segmentation system of inflectional affixes. In: CIPS-SIGHAN Joint Conference on Chinese Language Processing, Beijing, pp. 183–190 (2010)
7. Kairakbay, B.: A nominal paradigm of the Kazakh language. In: 11th International Conference on Finite State Methods and Natural Language Processing, St. Andrews, pp. 108–112 (2013)
8. Koehn, P., Hoang, H., Birch, A., Callison-Burch, C., Federico, M., Bertoldi, N., Cowan, B., Shen, W., Moran, C., Zens, R., Dyer, C., Bojar, O., Constantin, A., Herbst, E.: Moses: open source toolkit for statistical machine translation. In: 45th Annual Meeting of the Association for Computational Linguistics, Prague, pp. 177–18 (2007)
9. Tapias, D., Rosner, M., Piperidis, S., Odjik, J., Mariani, J., Maegaard, B., Choukri, K., Calzolari, N.: MultiUN: a multilingual corpus from united nation documents. In: Seventh conference on International Language Resources and Evaluation, La Valletta, pp. 868–872 (2010)
10. Moore, R.: Improving IBM word alignment model 1. In: 42nd Annual Meeting on Association for Computational Linguistics, Barcelona, pp. 518–525 (2004)
11. Och, F.J., Ney, H.: A Systematic Comparison of Various Statistical Alignment Models. Computational Linguistics **29**, 19–51 (2003)
12. Brown, P.F., Della-Pietra, V., Del-Pietra, S., Mercer, R.L.: The mathematics of statistical machine translation: Parameter estimation. Computational Linguistics **19**, 263–311 (1993)
13. Lindén, K., Axelson, E., Hardwick, S., Pirinen, T.A., Silfverberg, M.: HFST—framework for compiling and applying morphologies. In: Mahlow, C., Piotrowski, M. (eds.) SFCM 2011. CCIS, vol. 100, pp. 67–85. Springer, Heidelberg (2011)
14. Gabrilovich, E., Markovitch, S.: Computing semantic relatedness using wikipedia-based explicit semantic analysis. In: 20th International Joint Conference on Artificial Intelligence, Hyderabad, pp. 1606–1611 (2007)

15. Dunning, T.: Accurate methods for the statistics of surprise and coincidence. Computational Linguistics **19**, 61–64 (1993)
16. Papineni, K., Roukos, S., Ward, T., Zhu, W.: BLEU: a method for automatic evaluation of machine translation. In: 40th Annual Meeting of the Association for Computational Linguistics, Philadephia, pp. 311–318 (2002)
17. Och, F.J.: Minimum error rate training in statistical machine translation. In: 41st Annual Meeting of the Association for Computational Linguistics, Sapporo, pp. 160–167 (2003)
18. Federico, M., Bertoldi, N., Cettolo, M.: IRSTLM: an open source toolkit for handling large scale language models. In: Interspeech 2008, Brisbane, pp. 1618–1621 (2008)
19. Heafield, K.: Kenlm: faster and smaller language model queries. In: Sixth Workshop on Statistical Machine Translation, Edinburgh, pp. 187–197 (2011)
20. Clark, J.H., Dyer, C., Lavie, A., Smith, N.A.: Better hypothesis testing for statistical machine translation: controlling for optimizer instability. In: 49th Annual Meeting of the Association for Computational Linguistics, Portland, pp. 176–181 (2011)
21. Snover, M., Dorr, B., Schwartz, R., Micciulla, L., Makhoul, J.: A study of translation edit rate with targeted human annotation. In: Association for Machine Translation in the Americas, Cambridge, pp. 223–231 (2006)
22. Denkowski, M., Lavie, A.: Meteor 1.3: automatic metric for reliable optimization and evaluation of machine translation systems. In: Workshop on Statistical Machine Translation EMNLP 2011, Edinburgh, pp. 85–91 (2011)

First Workshop on PErvasive WEb Technologies, Trends and Challenges (PEWET 2015)

Fractally-Organized Connectionist Networks: Conjectures and Preliminary Results

Vincenzo De Florio$^{(\boxtimes)}$

MOSAIC/University of Antwerp and MOSAIC/iMinds Research Institute,
Middelheimlaan 1, 2020 Antwerp, Belgium
vincenzo.deflorio@gmail.com

Abstract. A strict interpretation of connectionism mandates complex networks of simple components. The question here is, is this simplicity to be interpreted in absolute terms? I conjecture that absolute simplicity might not be an essential attribute of connectionism, and that it may be effectively exchanged with a requirement for relative simplicity, namely simplicity with respect to the current organizational level. In this paper I provide some elements to the analysis of the above question. In particular I conjecture that fractally organized connectionist networks may provide a convenient means to achieve what Leibniz calls an "art of complication", namely an effective way to encapsulate complexity and practically extend the applicability of connectionism to domains such as sociotechnical system modeling and design. Preliminary evidence to my claim is brought by considering the design of the software architecture designed for the telemonitoring service of Flemish project "Little Sister".

1 Introduction

Connectionism—also known as parallel distributed processing (PDP) and artificial neural networks (ANN)—has been successfully applied to several problems, including pattern and object recognition, speaker identification, face processing, image restoration, medical diagnosis, and others [1], as well as to several cognitive functions [2]. In connectionism,

> "Processing is characterized by patterns of activation across **simple** processing units connected together into complex networks. Knowledge is stored in the strength of the connections between units." [2]

The accent on simplicity is also present in another definition of connectionism:

> "The emergent processes of interconnected networks of **simple** units" [3].

Similarly, in [4] Rumelhart, Hinton, and McClelland introduce PDP as a model based on a set of **small, feature-like** processing units.

I believe it is important to reflect on the *simplicity* requirement expressed in the above definitions. Regardless of their position and role, the nodes of a connectionist network are intended as simple parts. This is the case also when

© Springer International Publishing Switzerland 2015
F. Daniel and O. Diaz (Eds.): ICWE 2015 Workshops, LNCS 9396, pp. 53–64, 2015.
DOI: 10.1007/978-3-319-24800-4_5

the network is organized into a complex hierarchy of layers. Simplicity pertains to the function of the role but also to the role played by the node, which is tuneable though statically defined.

My question here is: is this simplicity to be interpreted *in absolute terms*? If that would be the case, then individual nodes may not represent complex behaviors resulting from the collective action of aggregations of other nodes. In this sense, absolute simplicity of the nodes may limit the applicability of connectionism. How could one easily and comfortably model, e.g., a complex social organization, or a digital ecosystem, or a biological organism, only by reasoning in terms of simple nodes? Such an endeavor would be the equivalent to writing a complex software application with no mechanism to encapsulate complexity (such as software modules, services, aspects, and components).

The rest of this article is to detail the reasons why my answer to the above question is "no". In fact, my conjecture is that absolute simplicity might not be an *essential* attribute of connectionism, and that it may be effectively exchanged with a requirement for **relative simplicity**, namely simplicity with respect to the current organizational level.

A second conjecture here is that a convenient hybrid form of connectionism would be that of *fractally organized connectionist networks* (FOCN). More formally, **fractal connectionism** would replace the absolute simplicity requirement of "pure" connectionism with the following two properties:

Fractal Organization: FOCN nodes are fractally organized [5]. This in particular means that nodes have a dynamic organizational role that depends on the context and on system-wide organizational rules—the so-called "canon". In other words, regardless of their level in a fractal hierarchy, the nodes obey the same canon and switch between, e.g., management and subordinate, or input and output roles, depending on the situation at hand. The nodes become thus *organizationally homogeneous*. In fractal organizations nodes are typically called holons [5] or fractals.

Modularity and Relative Simplicity: FOCN nodes function as modules that are at the same time monadic (namely, atomic and indivisible) with respect to the layer they reside in and composite organizations of parts residing in lower layers [6]. Through these "organizational digits" *absolute simplicity becomes relative simplicity*. As nonterminal symbols in context-free grammars, every node in FOCN is in itself both a network and the "root" of that network.

In what follows I provide some elements towards a discussion of the benefits of coupling fractal organization with connectionism.

- In Sect. 2 I first identify in the so-called Art of Complication of Leibniz the ancestor of "relative simplicity" and fractal organization.
- In Sect. 3 I briefly recall the major aspects of fractal social organizations, an organizational model for sociotechnical systems and cyber-physical societies. In particular in that section I compare the major differences between PDP and fractal social organizations. As a result of that comparison, fractal social organizations are interpreted as a FOCN organizational model.

- A practical application of said model is the subject of Sect. 4: the web service architecture and middleware developed in the framework of Flemish project "Little Sister".
- Conclusions and next steps are then drawn in Sect. 5.

2 Leibniz' Art of Complication

> When the tables of categories of our *art of complication* have been formed, *something greater will emerge*. For let the first terms, of the combination of which all others consist, be designated by signs; these signs will be a kind of alphabet. It will be convenient for the signs to be as natural as possible—e.g., for one, a point; for numbers, points; for the relations of one entity with another, lines; for the variation of angles and of extremities in lines, kinds of relations. If these are correctly and ingeniously established, this universal writing will be as easy as it is common, and will be capable of being read without any dictionary; at the same time, a fundamental knowledge of all things will be obtained. The whole of such a writing will be made of geometrical figures, as it were, and of a kind of pictures—just as the ancient Egyptians did, and the Chinese do today. Their pictures, however, are not reduced to a fixed alphabet [...] with the result that a tremendous strain on the memory is necessary, which is the contrary of what we propose. [7]
>
> *Of the art of combination*
> G. W. VON LEIBNIZ

As brilliantly discussed in [6], hierarchies are a well-known and consolidated concept that pervades the organization of our societies and that of biological and computer-based systems. Particularly interesting and relevant to the present discussion are so-called **nested compositional hierarchies**, defined in the cited reference as "a pattern of relationship among entities based on the *principle of increasing inclusiveness*, so that entities at one level are composed of parts at lower levels and are themselves nested within more extensive entities". As mentioned in Sect. 1, increasing inclusiveness (I^2) practically realizes modularity and relative simplicity by creating matryoshka-like concepts that are at the same time monadic and composite. The same principle and the same duality may be found in the philosophy of Leibniz [8,9]. The Great One introduces the concept of substances, namely

> "networks of other substances, together with their relationships. [... A substance is a] concept-network packaging a quantum of knowledge that becomes a new *digit*: a new concept so unitary and indivisible as to admit a new pictorial representation—a new and unique [pictogram]" [8].

Leibnitian pictograms were thus an application of the I^2 principle to knowledge representation. Pictograms of substances are thus at the same time knowledge units and knowledge networks; unique digits and assemblies of lower level signs and pictograms; which are obtained through some well-formed method of composition—some compositional *grammar*. Leibniz called the corresponding *language* "Characteristica Universalis" (CU): a knowledge representation language in which any conceptual model would have been expressed and reasoned upon in a mechanical way. The "engine", or algebra, for crunching CU expressions was called by Leibniz Calculus Ratiocinator. A parser reducing a sentence in a context-free language into a nonterminal symbol is a natural example of a Calculus Ratiocinator. As mentioned in [8],

> "Such pictograms represent modules, namely knowledge components packaging other ancillary knowledge components. In other words, pictograms are Leibniz's equivalent of Lovelace's and Turing's tables of instructions; of subroutines in programming languages; of boxes in a flowchart; of components in component-based software engineering."

Interestingly enough, the same principle and ideas were recently re-introduced in Actor-Network Theory [10,11] through the concepts of *punctualization* and *blackboxing*.

3   Connectionism vs. Fractal Social Organizations

Fractal social organizations (FSO) are a class of socio-technical systems introduced in [12,13]. FSO may be concisely described as a fractal organization of nodes called service-oriented communities (SoC's) [14]. Such nodes are "organizationally homogeneous", meaning that they provide the same, relatively simple organizational functions regardless of their place in the FSO network. Each node is a fractal—in the sense specified in Sect. 1—and may include other nodes, thus creating a matryoshka-like structure. A special node withing each SoC *punctualizes* the whole SoC. Such node is called representative and is at the same time both a node of the current SoC and a node of the "higher-ups"—namely the SoC (or SoC's) that include the current SoC. Nodes publish information and they offer and require services. Information and service descriptions reach the representative and are stored into a local registry. The arrival of new information and service descriptions triggers the execution of response activities, namely guarded actions that are enabled by the availability of data and roles. Missing roles triggers so-called exceptions: the request is forwarded to the higher-ups and the missing roles are sought there. Chains of exceptions propagate the request throughout the FSO and result in the definition of new temporary SoC's whose aim is executing the response activities. The lifespan of the temporary SoC's is limited to the execution of the activities they are associated with. Due to the exception mechanism, the new temporary SoC's may include nodes that belong to several and possibly "distant" SoC's. As such they represent an overlay network that is cast upon the FSO. Because of this I call them "social overlay networks" (SON).

In order to assess the relationship between FSO and PDP, now I briefly review the components of the PDP model as introduced by Rumelhart, Hinton, and McClelland in [4]. For each component I highlight similarities and "differentiae" [15], namely specific differences with respect to elements of the FSO model [16].

In what follows, uncited quotes are to be assumed from [4].

Fig. 1. Space of all possible states of activation of an FSO with nine agents and six roles. Roles are represented as integers $0, \ldots 6$. Role 0 is played by four agents, all the other roles are played by one agent.

- "A set of processing units". In PDP these units may represent "features, letters, words, or concepts", or they may be "abstract elements over which meaningful patterns can be defined". Conversely, in FSO those units are *actors*, identified by a set of integers [12]. A major difference is that in FSO actors can be "small, feature-like entities" but also complex collections thereof. Another peculiarity of FSO is given by the presence of a special role—the above mentioned representative.
- "A state of activation". In PDP this refers to the range of states the processing nodes may assume over time. In PDP this range may be discrete or continuous. A simple example is given by range $\{0, 1\}$, interpreted as "node is inactive" and "node is active".

  In FSO the state of activation is simply whether an actor is involved in an activity and thus is playing a role, or if it is inactive. In [12] I described the global state of activation of FSO by means of two dynamic systems, $L(t)$ and $R(t)$, respectively representing all inactive and all active FSO actors at time t. Pictures such as the one in Fig. 1 represent the space of all possible states of activation of an FSO. Actors can request services or provide services—which corresponds to the input and output units in [4]. The visibility of actors is restricted by the FSO concept of *community*: a set of actors in physical or logical proximity, for the sake of simplicity interpreted as a *locus* (for instance a room; or a building; or a city, etc). Non-visible actors correspond to the hidden units of PDP [4].
- The behaviors produced by the activated actors of an FSO correspond to what Rumelhart et al. call as the "output of the units" in PDP. In FSO, this behavior is cooperative and is mediated by the representative node. In the current implementation of our FSO models, a node's output is equal to the state of activation. In other words, in FSO an actor is currently either totally involved in playing a role or not at all. Future, more realistic implementations will introduce a percentage of involvement, corresponding to PDP's unit output. This will make it possible to model involvement of the same actor in multiple activities.
- Nodes of a PDP network are also characterized by a "pattern of connectivity", namely the interdependencies among the nodes. Each PDP node, say node n, has a *fan-in* and a *fan-out*, respectively meaning the number of nodes that may have an influence on n and the number of nodes that may be influenced by n. Influence has a *sign*, meaning that the corresponding action may be either excitatory or inhibitory. Conversely, in FSO I distinguish two phases—construction and reaction. In *construction*, the only pattern of connectivity is between the nodes of an SoC and the SoC representative. This pattern extends beyond the originating SoC by means of the mechanism of exception and results in the definition of a new temporary SoC—the already mentioned SON. Once this is done, *reaction* takes place with the enaction of all the SON agents. Different patterns of activity may emerge at this point, representing how each SON agent contributes to the emerging collective behavior of the SON.

– Another element of the PDP model is the "rule of propagation", stating how "patterns of activities [propagate] through the network of connectivities" in response to an input condition. In FSO, propagation is simply regulated by the canon, namely the rules of the representative and of the exception [16].
– So-called "activation rule" is a function modeling the next state of activation given the current one and the state of the network. In the current model, FSO activation rules are very simple and dictate that any request for enrollment to an inactive role is answered positively. A more realistic implementation should model the propensity and condition of a node to accept a request for enrollment in an activity. Factors such as the availability of the node, its current output level (namely, degree of involvement), and even economic considerations such as intervention policies and "fares" should be integrated into our current FSO model.
– Another important component of the PDP model is "Modifying patterns of connectivity as a function of experience". As suggested in our main reference, "this can involve three types of modifications:
  • The development of new connections.
  • The loss of existing connections.
  • The modification of the strengths of connections that already exist."
As mentioned above, in FSO we have two types of connections:
1. "Institutional" connections, represented by relationships between organizationally stable SoC's. An example is a "room" SoC that is stably a part of a "smart house" SoC, in turn a stable member of a "smart building" SoC.
2. "Transitional" connections, namely connections between existing SoC's and new SON's.
As I suggested in [12], experience may play an important role in FSO too. By tracking the "performance" of individual nodes (as described, e.g., in [17,18]) and individual SON's the structure and processing of an FSO may *evolve*. In particular, transient SON may be recognized as providing a recurring "useful" function, and could be "permanentified" (namely, turned into a new permanent SoC). An example of this may be that of a so-called "shadow responder" [19] providing consistently valuable support in the course of a crisis management action. Permanentification would mean that the shadow responder—for instance, a team of citizens assembled spontaneously and providing help and assistance to the victims of a natural disaster—would be officially or de facto recognized and integrated in the "institutional" response organizations, as suggested in [20,21].

Similarly, SoC that repeatedly fail to provide an effective answer to experienced situations may cease to make sense and be removed from the system. Reorganizations are a typical example of cases in which this phenomenon may occur.

The PDP concept of the strength of connection is also both interesting and relevant to the present discussion. A connection between two nodes may be realized as being "mutually satisfactory" (what is sometimes called as a "win-win") and in the long run may strengthen by producing a stable connection. Mutualistic relationships such as symbiosis and commensalism are typical examples of this phenomenon. Their role in FSO has been highlighted in [22].

– A final element in PDP is the "representation of the environment". This is a key component in the FSO model too, though with a very different interpretation of what an environment is. In PDP environment is "a time-varying stochastic function over the space of input patterns", while in FSO it is the set of probabilistic distributions representing the occurrence of the input events. As an example, environment is interpreted in FSO also as the rate at which new requests for assistance enter the FSO.

4   The Little Sister Software Architecture

Little Sister (LS) is the name of a Flemish ICON project financed by the iMinds research institute and the Flemish Government Agency for Innovation by Science and Technology. The project run in 2013 and 2014 and aimed to deliver a low-cost telemonitoring [23] solution for home care. Two are the reasons for mentioning LS here:

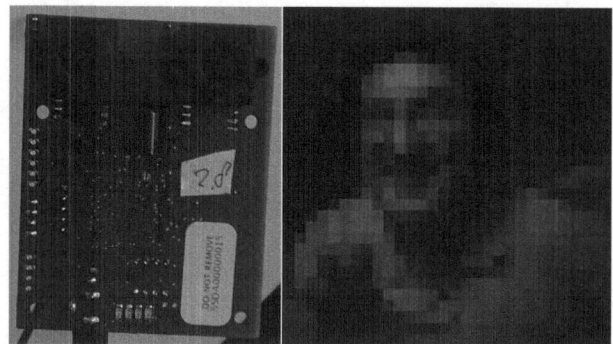

Fig. 2. The Little Sister mouse-cam sensor [24] and an exemplary picture taken with it.

1. LS may be considered as a connectionist approach to telemonitoring: in fact in LS the collective action of an interconnected network of simple units [24] (battery-powered mouse sensors) replaces the adoption of more powerful and expensive complex devices (smart cameras; see Fig. 2).
2. The LS software architecture realizes a simplified FSO: a predefined set of SoC's realizes the structure exemplified in Fig. 3.

The cornerstone of the LS software architecture is given by web services standards. As discussed in [13],

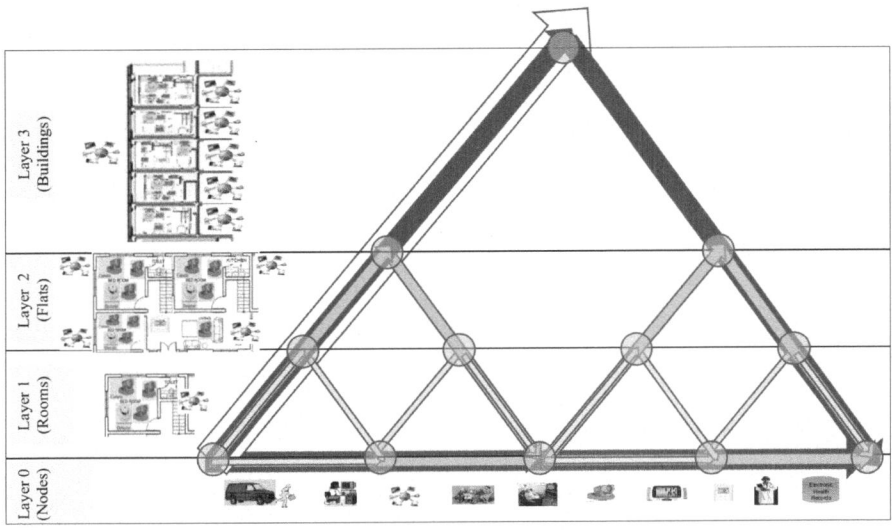

Fig. 3. Exemplification of the LS Fractal Social Organization.

"the LS mouse sensors are individually wrapped and exposed as manageable web services. These services are then structured within a hierarchical federation reflecting the architectural structure of the building in which they are deployed [25]. More specifically, the system maintains dedicated, manageable service groups for each room in the building, each of which contains references to the web service endpoint of the underlying sensors (as depicted in layers 0 and 1 in Fig. 3). These 'room groups' are then aggregated into service groups representative of individual housing units. Finally, at the highest level of the federation, all units pertaining to a specific building are again exposed as a single resource (layer 3). All services and devices situated at layers 0–3 are deployed and placed within the building and its housing units; all services are exposed as manageable web services and allow for remote reconfiguration."

Absolute simplicity is here traded with modularity and relative simplicity: each "level" hosts nodes that are "simple" with respect to the granularity of the action. Correspondingly, each layer hosts services of increasing complexity, ranging from image to motion processing and from raw context perception to situation identification [26]. Each SoC is managed by a representative implemented as a module of a middleware. Said middleware is based on a fork of Apache MUSE—"a Java-based implementation of the WS-ResourceFramework (WSRF), WS-BaseNotification (WSN), and WS-DistributedManagement (WSDM) specifications" [27] on top of Axis2 [28], and partially implements the WSDM-MOWS specification [29] (Web Services Distributed Management: Management of Web Services).

It is the LS middleware component in each SoC that manages the FSO canon: events produced by the local nodes are received by the middleware by means of a standardized, asynchronous publish-and-subscribe mechanism [30]. The middleware then verifies whether any of the local nodes may respond with some actuation logic. If so, the local node is appointed to the management of the response; otherwise, an exception takes place (see Sect. 3) and the event is propagated to the higher-up SoC. Given the fact that, in LS, a predefined population of nodes and services is available and known beforehand, the selection and exception mechanisms are simple and have been implemented by annotating events and services with topic identifiers. In a more general implementation of the FSO model, selection and exception require semantic description and matching support as discussed in [31].

5   Conclusions

> At a low level of ambition but with a high degree of confidence [General Systems Theory] aims to point out similarities in the theoretical constructions of different disciplines, where these exist, and to develop theoretical models having applicability to at least two different fields of study. At a higher level of ambition, but with perhaps a lower degree of confidence it hopes to develop something like a "spectrum" of theories—a system of systems which may perform the function of a "gestalt" in theoretical construction. Such "gestalts" in special fields have been of great value in directing research towards the gaps which they reveal. [...] Similarly a "system of systems" might be of value in directing the attention of theorists toward gaps in theoretical models, and might even be of value in pointing towards methods of filling them.
>
> *General Systems Theory—The Skeleton of Science*
> K. BOULDING

In this work I considered two seemingly unrelated "gestalts": connectionism and fractal organization. By reasoning about them in general and abstract terms, I observed how connectionism could possibly benefit from the application of I^2, namely the principle of increasing inclusiveness, and by interpreting processing nodes' simplicity in relative rather than absolute terms. I have conjectured that, in so doing, connectionism would further extend its applicability and expressiveness. I called fractally-organized connectionist networks the resulting hybrid formulation. I then introduced a model of fractal organization called FSO and I compared the key elements of parallel distributed processing with corresponding assumptions and strategies in FSO. As a practical example of the hybrid model I discussed the software architecture of Flemish project "Little Sister"—a web services-based implementation of a "fractally connectionist" system. As observed by Boulding [32], the above discussion put to the foreground a number of oversimplifications in the current FSO model. As a consequence, our future research shall be *directed towards the gaps that the above discussion helped revealing*, in particular extending the FSO model with more complete and general elements of the connectionist models.

Acknowledgment. This work was partially supported by iMinds—Interdisciplinary institute for Technology, a research institute funded by the Flemish Government—as well as by the Flemish Government Agency for Innovation by Science and Technology (IWT). The iMinds Little Sister project is a project co-funded by iMinds with project support of IWT (Interdisciplinary institute for Technology) Partners involved in the project are Universiteit Antwerpen, Vrije Universiteit Brussel, Universiteit Gent, Xetal, Niko Projects, JF Oceans BVBA, SBD NV, and Christelijke Mutualiteit vzw.

References

1. Mira, J., Prieto, A.G. (eds.): IWANN 2001. LNCS, vol. 2085. Springer, Heidelberg (2001)
2. Thomas, M.S., McClelland, J.L.: Connectionist models of cognition. In: Sun, R. (ed.) The Cambridge Handbook of Computational Psychology. Cambridge University Press (2008)
3. Wittek, P.: Quantum Machine Learning: What Quantum Computing Means to Data Mining. Academic Press, San Diego (2014)
4. Rumelhart, D.E., Hinton, G.E., McClelland, J.L.: A general framework for parallel distributed processing. In: Rumelhart, D.E., McClelland, J.L., PDP Research Group, C. (eds.) Parallel Distributed Processing: Explorations in the Microstructure of Cognition, vol. 1, pp. 45–76. MIT Press, Cambridge (1986)
5. Koestler, A.: The Ghost in the Machine. Macmillan (1967)
6. Tëmkin, I., Eldredge, N.: Networks and hierarchies: approaching complexity in evolutionary theory. In: Serrelli, E., Gontier, N. (eds.) Macroevolution: Explanation, Interpretation, Evidence. Springer (2014)
7. Leibniz, G.W.: Of the art of combination (1666). In: Leibniz: Logical Papers. Clarendon Press, Oxford, (1966) English translation of a portion of Chapter 11 by G.H.R. Parkinson
8. De Florio, V.: Behavior, organization, substance: three gestalts of general systems theory. In: Proceedings of the IEEE 2014 Conference on Norbert Wiener in the 21st Century. IEEE (2014)
9. De Florio, V.: Systems, resilience, and organization: Analogies and points of contact with hierarchy theory (2014). CoRR abs/1411.0092
10. Latour, B.: Pandora's hope: essays on the reality of science studies. Harvard University Press, Cambridge (1999)
11. Latour, B.: On actor-network theory. a few clarifications plus more than a few complications. Soziale Welt **47**, 369–381 (1996)
12. De Florio, V., Bakhouya, M., Coronato, A., Di Marzo Serugendo, G.: Models and concepts for socio-technical complex systems: Towards fractal social organizations. Systems Research and Behavioral Science **30**(6) (2013)
13. De Florio, V., Sun, H., Buys, J., Blondia, C.: On the impact of fractal organization on the performance of socio-technical systems. In: Proceedings of the 2013 International Workshop on Intelligent Techniques for Ubiquitous Systems (ITUS 2013). IEEE, Vietri sul Mare (2013)

14. De Florio, V., Blondia, C.: Service-oriented communities: visions and contributions towards social organizations. In: Meersman, R., Dillon, T., Herrero, P. (eds.) OTM 2010. LNCS, vol. 6428, pp. 319–328. Springer, Heidelberg (2010). doi:10.1007/978-3-642-16961-8_51

15. Burek, P.: Adoption of the classical theory of definition to ontology modeling. In: Bussler, C.J., Fensel, D. (eds.) AIMSA 2004. LNCS (LNAI), vol. 3192, pp. 1–10. Springer, Heidelberg (2004)

16. De Florio, V., Pajaziti, A.: How resilient are our societies? analyses, models, and preliminary results (2015). CoRR abs/1505.02759

17. Buys, J., De Florio, V., Blondia, C.: Towards parsimonious resource allocation in context-aware n-version programming. In: Proceedings of the 7th IET System Safety Conference, The Institute of Engineering and Technology (2012)

18. Buys, J., De Florio, V., Blondia, C.: Towards context-aware adaptive fault tolerance in SOA applications. In: Proceedings of the 5th ACM International Conference on Distributed Event-Based Systems (DEBS), pp. 63–74. Association for Computing Machinery, Inc. (ACM) (2011)

19. De Florio, V., Sun, H., Blondia, C.: Community resilience engineering: reflections and preliminary contributions. In: Majzik, I., Vieira, M. (eds.) SERENE 2014. LNCS, vol. 8785, pp. 1–8. Springer, Heidelberg (2014)

20. RAND: Community resilience (2014). http://www.rand.org/topics/community-resilience.html

21. Colten, C.E., Kates, R.W., Laska, S.B.: Community resilience: Lessons from new orleans and hurricane katrina. Technical Report 3, Community and Regional Resilience Institute (CARRI) (2008)

22. De Florio, V., Sun, H., Bakhouya, M.: Mutualistic relationships in service-oriented communities and fractal social organizations (2014). CoRR abs/1408.7016

23. Meystre, S.: The current state of telemonitoring: a comment on the literature. Telemed J. E-Health **11**(1), 63–69 (2005)

24. Anonymous: Introducing the Silicam IGO (2013)

25. OASIS: Web services service group 1.2 standard. Tech. report, OASIS (2006)

26. Ye, J., Dobson, S., McKeever, S.: Situation identification techniques in pervasive computing: A review. Pervasive and Mobile Computing **8**(1), 36–66 (2012)

27. Anonymous: Apache Muse—a Java-based implementation of WSRF 1.2, WSN 1.3, and WSDM 1.1 (2010). http://ws.apache.org/muse (retrieved on June 11, 2015)

28. Anonymous: Apache Axis2/Java—next generation web services (2010). http://ws.apache.org/axis2 (retrieved on June 11, 2015)

29. Anonymous: Web services distributed management: Management of web services (WSDM-MOWS) 1.0 (2004). http://xml.coverpages.org/WSDM-CD-10971-MOWS10.pdf (retrieved on May 17, 2015)

30. OASIS: Web services base notification 1.3 standard. Tech. rep., OASIS (2006)

31. Sun, H., De Florio, V., Blondia, C.: Implementing a role based mutual assistance community with semantic service description and matching. In: Proc. of the Int.l Conference on Management of Emergent Digital EcoSystems (MEDES) (2013)

32. Boulding, K.: General systems theory–the skeleton of science. Management Science **2**(3) (1956)

Internet-Based Enterprise Innovation Through a Community-Based API Builder to Manage APIs

Romanos Tsouroplis$^{(\boxtimes)}$, Michael Petychakis, Iosif Alvertis, Evmorfia Biliri, Fenareti Lampathaki, and Dimitris Askounis

School of Electrical and Computer Engineering,
National Technical University of Athens, Athens, Greece
{rtsouroplis,mpetyx,alvertisjo,ebiliri,flamp,askous}@epu.ntua.gr

Abstract. More and more users, these days, keep fragmented data across the web in different applications, through various types of devices, PC, mobiles, wearable devices, etc. By taking advantage of an aggregative Graph Application Programming Interface (API), users have the ability to harness shattered data and keep them into a privacy-aware platform (Cloudlet) where permissions can be applied, and therefore let developers build useful applications from it. To make this unifying Graph API, the API Builder is proposed, as a tool for easily creating new APIs and connecting them with existing ones from Cloud-based Services (CBS), thus providing integration among services and making it easier for users and/or enterprises to reach a larger audience while conveying their message. Typical obstacles, like keeping up to date with CBS API versioning, that seems daunting for developers, are also tackled through semi-automation and the help of the community empowering the API Builder. In that way, application developers do not have to worry for merging various APIs or if the application-generated data are locked in silos of companies; now the user is the judge who gives access to their data and meta-data (i.e. especially context), to enable smarter, context-adaptive and richer in content applications.

Keywords: APIs · Cloud-based services · Evolving APIs · API Builder · Graph API · Community-based platform

1 Introduction

Social media applications and mobile devices have facilitated the constant, ubiquitous creation of information by users, and their needs to consume more content and communicate in various ways have allowed new business models and companies to grow [1]. Such favored, by this social and mobile era, companies and initiatives have built silos accessible through restricted APIs, either to protect users' privacy or their own business. It is desirable to protect users against malicious attacks, but not allowing them to transfer, see and control their data in need is also worth-mentioned. Apart from a possible value loss for users, what users lose at the end is the capability of additional value creation through new software solutions, which build on existing information and provide new solutions; users are tired to generate duplicate of infor-

© Springer International Publishing Switzerland 2015
F. Daniel and O. Diaz (Eds.): ICWE 2015 Workshops, LNCS 9396, pp. 65–76, 2015.
DOI: 10.1007/978-3-319-24800-4_6

mation, and want their applications to "just work", thus it is "unfair" for organizations to lock them in and leave unexplored data that could boost pervasive computing to new levels.

Web and data standards have evolved in such a level that it is unjustified not to have a unified API-based web. Both companies and users can be benefited by open innovation paradigms [2, 3] and find new solutions based on the plethora of data and meta-data located now in such large silos; just think how Google, Facebook or Twitter have built impressive businesses based on data algorithms on personal data, to fuel their business model with targeted and more effective advertising. Thus competition is currently not based on technological, engineering or design excellence, but on first-mover advantage and unfair competition.

Some may claim that the Web is not yet ready to expose common APIs to services, through trusted and well-designed interfaces. Nevertheless, just in the relative service for API documentation, Programmable Web, someone can find indexed, categorized and used on mashups about 12,987 public APIs listed, a number which is continuously growing [4]. Some others may claim that there are no standards yet to drive an open approach on web development; even if standards like Schema.org or best practices in semantic APIs like the DBpedia site, it is a constant headache for a developer, indeed, to read different documentations and build separate connectors for every single one service that they want to use, or keep in track with constant changes in the API landscape. But it should be easy for developers to easily connect to every cloud-based service, in order to exchange semantically clear data, and put emphasis on data analysis, in order to make pervasive powerful and independent of initial training or user calibration.

Another problem for developers, trying to connect and build over API platforms, is the constant changes common cloud-based services (CBS) undergo, even in a yearly basis; APIs evolve and change as business needs change and policies are altered. For example Facebook API, one of the most popular APIs for third party developers, is by the time this paper is written on version 2.2, and all applications using version 1.x of the API are discontinued after April 30th 2015 [5]; even if developers are notified in advance, many times changes in APIs are massive, like the integration of check-in object into the status object of Facebook in second version. Some platforms try to overcome such headaches, by ensuring developers for supporting previous versions, like Dropbox does [6], but such a strategy applies only on businesses where APIs can remain static and policies do not change frequently. All these difficulties are also stated in the relative work of Tiego Espinha et al. [7].

Many studies try to measure the impact and identify the effects of all these changes in APIs, both in third party developers [8, 9] and in changes needed to be undertaken in back-end services, with code refactoring [10]. In each case, the client cannot easily be aware of changes that take place in the back-end, and adapt automatically client's operations on changes on APIs on servers.

OPENi has already worked to address problems in modeling APIs, under common standards and a user-centric, graph design, called Generic APIs and applying in various categories (i.e. shopping, media, activity API) [11, 12]. This paper extends this work and presents how developers can collaborate to extend these objects, APIs or properties under the OPENi API Builder, in order to extend their applications capabilities and improve the capabilities of their applications for pervasive computing, without worrying for proper modeling and standards. In addition, as OPENi API Builder has mapped Generic APIs with various CBS methods and objects, developers can (a) develop applications connected with common platforms with minor effort, and (b) maintain and support only one interface to connect with APIs that change versions through time; in that way, developers have to learn and use only one API, and the problem of maintaining connectivity with existing CBS APIs has moved to a prospective community of developers who collaborate - like in Wikipedia - in order to keep up-to-date; the effectiveness of such an approach has still to be observed and evaluated

The structure of the present paper is as follows. Section 2 gives the background and the related work in the area, section 3 presents the methodology followed towards the implementation of the OPENi API Builder. In section 4, an overview of the API Builder is given, including the added value features provided by the API Builder, broken up into three smaller sections, the community-based orientation filled with social characteristics, the extensive documentation in all the standards currently available and lastly the case of CBS API changes handling. Section 5 concludes the paper with future work on this matter.

2   Background and Related Work

The field of APIs and API creation has gained a lot of attention in the last years. As more developers and enterprises rely on APIs to communicate and exchange information internally or outside of an application, more tools are created to make this process easier, as standalone software products or open source solutions.

An essential part of having an API is the provided documentation, either inside or outside of a team. Helpful and easy to use documentation can lead to increased productivity, less questions and frustrated people as well as easier maintenance of the back-end code. Despite this fact, many APIs are not accompanied with well-structured documentation, while some do not provide one at all. Another common problem when a new technology or methodology is adopted is the lack of standards in the first years. That is not quite the case with the APIs, but still a large portion of them do not conform to a particular standard.

A number of REST metadata formats has been created for documenting APIs. A common documentation scheme includes the methods that can be used (GET, POST, PUT, DELETE, etc.) along with the resource parameters, required or optional, description of what these methods do, testing functionality and status/error codes which clarify how requests are processed and what the retrieved responses mean.

The most popular standards can be found in the following table:

- **API Blueprints.** An API description written in this format can be used in the apiary platform to create automated mock servers, validators etc. [13]

- **RAML.** A well-structured and modern API modelling language, gaining ground daily [14]
- **Hydra.** The Hydra specification [15] is currently under heavy development. Its added value is that it tries to enrich current web APIs with tools and techniques from the semantic web area to create sustainable services in the long term.
- **WADL.** This XML based specification was submitted to W3C, but did not succeed in being widely adopted [16]
- **Swagger.** Offers a large ecosystem of API tooling, has very large community of active users and great support in almost every modern programming language and deployment environment. [17]

Table 1. API Specification Details

Details\Specs	API Blueprints	RAML	Hydra	WADL	Swagger
Format	Markdown	YAML	Hydra Core Voc. - JSON-LD	XML	JSON
Licence	MIT	ASL 2.0/TM	CC Atrribution 4.0	Sun Microsystems Copyright	ASL 2.0
Available	Github	Github	www.w3.org	www.w3.org	Github
Sponsored By	Apiary	Mulesoft	W3C Com Group	Sun Microsystems	Reverb
Version	Format 1A revision 7	1.0	Unofficial Draft 19 January 2015	31 August 2009	2.0
Initial Commit	April 2013	Sep 2013	N/A	Nov 2006	July 2011
Pricing Plans	Y	Y	N	N	N
StackOverflow Questions	75	37	35	156	732
Github Stars	1819	1058	N/A	N/A	2459
Google Search	1.16M	457k	86k	94.1k	28M

Apart from the provided documentation, there are best practices based on standards as well, regarding the response given after a request is made to an API. The most prevalent format specifications are HAL [18], JSON API [19], JSON-LD [20] and Collection+JSON [21], with the two first getting the most attention. It is important to model the API responses to make sure that clients do not break their communication to the services and thus seamless business integrations can continue to flourish. Those approaches are in general efforts to model existing popular media types and create a consensus to the chaos of already numerous IANA registered media types.

In the last years, the amount of APIs created is growing exponentially. An important issue is how anyone could find the API needed in a more efficient and less time-consuming way. For that matter, various solutions have been proposed, like {API}Search [22], ProgrammableWeb [4] or Mashape [23]. The first one utilizes the APIS.JSON [24] format so as to create indexes of an API in its search machine. Mashape is considered a marketplace of APIs, providing documentation and code samples along with the search feature. It also offers monitoring and analytics services for the API providers. Monitoring and Testing functionality can also be found at Runscope [25], a tool for inspecting and debugging an API.

Furthermore, applications that automatically create Software Development Kits (SDK) exist to make the integration with the consumer-client software side easier. Apimatic [26] and REST United [27] are the most well-known of these tools. Automation is provided in terms of importing a Swagger documentation and of exporting SDKs in 8-9 programming languages for web and mobile-based software.

Several applications stand out as core API managing suites. These are described in more details below:

1. Apigility [28] is an open source tool created by Zend Framework, running its own hosting environment, has visual interface for creating APIs with the ability to produce documentation in Swagger.
2. StrongLoop Arc [29], which provides a nice user interface for implementing APIs that are then rendered via a Swagger-based API explorer. There is no community built upon this solution.
3. Apiary [13] uses API Blueprints code to provide the documentation along with additional features. The downside is that the API management is accomplished via coding. The Apiary solution has a community but no social characteristics.
4. API Designer by Mulesoft [30] has no visual interface as well. The developer writes RAML code to create an API.
5. Appcelerator [31], a company which targets mobile apps, gives an interface for rapid assembly and hosting of mobile-optimized APIs with some prebuilt connectors.
6. Apigee Edge API Platform [32] innovates in providing good analytic services for the APIs while providing most of the available CBS connectors
7. Kimono [33], by kimonolabs is a very interesting tool which allows users to scrape a website, by simply selecting data on the site. The tool then gathers similar information from the site into an endpoint for a new API. It does not have the same flexibility in managing your API but instead tries to automate the process in existing websites. No API documentation standard is yet provided.

Table 2. Related Work

	Features\Solutions	Apiary	API Designer	Apigee	Apigility	Appcelerator	kimono	StrongLoop Arc	API Builder
Specification	Swagger			tweaked ui					
	Hydra								
	RAML								
	WADL								
	API Blueprints								
	user interface								
	open source								
	community	team-oriented				team-oriented			
	social network								
	Datastore Connectors								
	Connectivity with CBS					static			dynamic
	Matching with CBS								

In the table above, API Documentation Specification presents the usage of the previously described standards of Swagger, Hydra, RAML, WADL and API Blueprints. 4 out of 8 solutions have used the Swagger Specification, while RAML and API Blueprints are used as well. Apigee is among the first ones supporting Swagger.

At the API Builder, all standards are integrated and Swagger user interface is used for interaction with the developers. To the best of our knowledge, only the API Builder supports all forms of standards for the documentation.

User Interface is considered to be the visual representation of the APIs management board with buttons, textboxes and labels. Apiary and API Designer do not have this feature as they need programming code for the APIs administration. The only open source alternative to the API Builder is Apigility. For most of the other solutions pricing plans are provided.

Apiary and Appcelerator provide a team-oriented solution for collaboration in building an API, whereas kimono and API Designer have a community orientation. The first three have social features like commenting, following and liking other APIs. Apiary enables these functionalities inside a team working on a project. Extra attention has been given to the OPENi API Builder on community orientation and social characteristics for getting feedback on projects.

Connectivity with Cloud-based Services (CBS) is predefined for Appcelerator for a range of enterprise and public data sources like SAP, Oracle, Facebook, Twitter, etc. This is one of the main advantages found in the API Builder where the Cloud-based Services can be altered dynamically through the community. Last but not least, the feature of matching Objects of APIs to other Objects from CBS can only be met at the API Builder. More details on this matter can be found at Section 4.

3   Methodology

While creating the Generic Objects for the Graph API Platform in the context of OPENi [11], it was realized that advanced machinery was needed to utilize the complicated mechanisms of such an API. Thus, the API Builder was conceived as the solution for this specific problem that was later generalized to cover broader aspects of any API lifecycle. The first issue that had to be dealt with, was the high complexity of managing and orchestrating the changes in all of the CBS and the second of equal importance, a mechanism that would satisfactory handle potential changes in the Generic Objects had to be implemented. Therefore, extensive research was conducted on the field of the API lifecycle, including the creation, management, update and usage. This was completed in 7 stages, based of course on the modeling undertaken already for the Graph API. Validation of each step and peer-reviewing of the work done helped throughout the whole process to keep constantly up-to-date.

- **Step 1. Identifying the Needs for Handling the Graph API Evolution.** At this point the Graph API Platform was taken apart to realize which segments would be needed to update occasionally. This includes the Generic APIs, their Objects, the methods implemented for each one of them and last but not least, the matching with the Cloud-based Services. In this step, user stories were created to describe the expected functionalities.
- **Step 2. Research of Related Work.** API creation tools can be found in many enterprise solutions across the Web, on studies and even as open source framework for specific code languages. All of these solutions were gathered and categorized based on the providing features.

- **Step 3. Connectivity with CBS.** It is already discussed in detail, that nowadays it is widely accepted most services are based on other Cloud-based Services in order to increase their reach on users as well their revenue. The categorization on the CBS domain has already taken place for the Graph API Platform [11]. Some of the researched alternatives have static CBS API connectivity while most have none.
- **Step 4. Automation of the Evolution Process.** The ways on how the evolution of APIs happen have already been discussed by Dig and Johnson [10], where they try to explore the various automation that accompany such a process. They do not provide a solution on that matter without examining the Cloud-based Services' back-end code and this seems like in general the common approach. Therefore, semi-automatic was the chosen approach with the involvement of a swift-responding to changes community equipped with the right tools.
- **Step 5. Building upon Standards.** All of the available solutions depend on one or none of the thriving standards on API documentation. Support on the most significant of them was decided after carefully exploring them and their impact on the developer community.
- **Step 6. Build Scalable Solution.** The Builder should be able to manage multiple simultaneous user actions and keep a massive amount of data for objects and interconnections with CBS, so an analysis of best practices towards this end was taken into consideration.
- **Step 7. Crafting UI Mockups** to better express the power of the Builder. After assembling the requirements for making the API Builder, a great user experience had to be achieved through an interface. The design principles followed included intuitiveness, familiarity, simplicity, availability and responsiveness. Seamless functionality without need of documentation, globally-identified terminology and symbols, focus on frequently performed tasks while not excluding the rest from being visible and feedback on each of the actions performed were only some of the distinct principles endorsed.

4   API Builder

The OPENi API Builder is a Web-based platform, implemented as an open source project, publicly available in Github [34]. It is designed and implemented from a developer's perspective, aiming to facilitate all tasks related to API management, spanning from semi-automatic creation to versioning policies and deprecation. By adopting the simplified Builder API management processes, enterprises can reduce the time their developers spend on tedious tasks, leading to more productive developing time and ultimately to increased profits.

As depicted in Figure 1 the Builder user interface is simple and clean, so that the user can easily explore and use it without being distracted. Internally, APIs are stored as complex structures that may contain one or more objects. In turn, each object contains properties and links to cloud based services (CBS) and is associated to the methods that can be applied on it.

Any authorized developer can navigate through the Builder APIs and test them, give feedback to their developer, use them in his own applications, extend and even update them, or just create his own APIs and objects, leveraging the provided recommendations that are provided during the creation process.

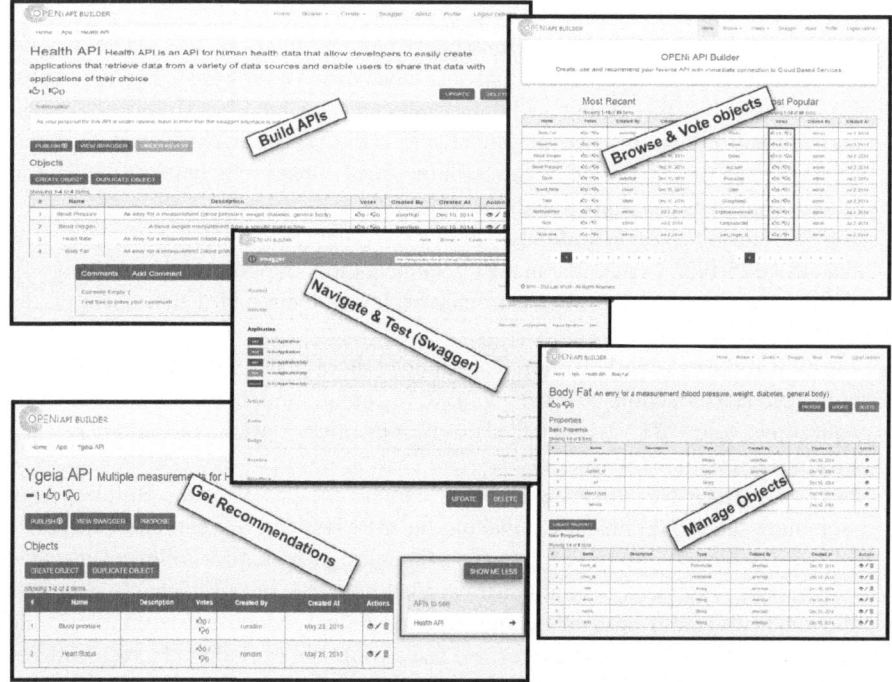

Fig. 1. API Builder Actions

The main innovations of the API Builder focus in the following three directions: community orientation, documentation based on standards, CBS changes handling which will be described in detail in the following sections.

Community Orientation

The advantages of transforming a traditional firm-based model into a community-based development process have been discussed in [35, 36]. The World Wide Web has enabled data and information exchange, from static content to the Web 2.0 revolution of dynamic user interactions and social bonds. Spanning from the Linux Kernel [37] to the Microsoft .NET tools and frameworks, open source solutions have proven to be effective in large user community creation and many modern business products aim to make the end-users interact and bind with them.

The OPENi API Builder aims to build such a community and satisfy all its needs in order to keep it active. Users are presented with a number of options regarding their interaction with the existing APIs and their components, i.e. objects, properties, methods and CBS. They can view, create, duplicate, update or delete any of the previously mentioned Builder parts, based on fine-grained permissions and strict workflows. Social characteristics, such as commenting, up-down voting and following other users or specific APIs are not only supported, but also encouraged through colorful links and buttons throughout the web interface. The social dimension is not just meant as a means to engage the developers in active participation, but hold a significant practical role as well. The votes and the number of followers of an API can be indicative of its usability and thus help guide the community towards better implemented solutions.

Although community incentives are usually strong defensive mechanism against malicious actions and spammers, explicit user roles and privileges are also designed in order to protect the developed APIs. In Table 3, the available user actions are put under the prism of API privacy option. An authenticated enterprise developer may select one of three available privacy options when creating a new API, namely: public, protected or private. Public APIs are open to anyone to extend and update, protected ones are viewable by anyone and can also be forked, while private are visible only to their owner.

Table 3. API Privacy

APIs	Authorized User Actions									
	Create	Duplicate	View	Update	Delete	Propose	Vote Up	Vote Down	Comment/Reply	Follow
Public	Y	Y	Y	Y	Y	Y	Y	Y	Y	Y
Protected	Y	Y	Y			Y	Y	Y	Y	Y
Private	Y									

In order to encourage the adoption of useful Objects and consequently avoid duplicate declarations, during the creation of a new object the developer is presented with recommendations of existing ones that could be used instead. These recommendations are calculated using approximate string matching on API, Object and property names as well as similarity computation on the provided descriptions, also on the property types.

After creating and publishing an API, an apis.json [24] file is created by using the information provided for the application, so that it can be indexed in the apis.io directory [22] and that it can be discovered by more interested parties.

Documentation Based on Standards

As is already discussed in the Background & Related Work section, there is no commonly adopted standard for describing Restful APIs. Even though Swagger is at the moment the leading alternative, in order to enable developers to become part of the Builder community, regardless of the format they prefer to use and broaden the pool of APIs that they can test and interact with, a format transformation service has been developed. The OPENi API Builder offers a mapping functionality, i.e. the ability to transform API descriptions amongst Swagger [17], RAML [14], API Blueprints [13], WADL [16] and Hydra [15] specifications.

An enterprise may expose its API to one of the aforementioned five specifications and any interested party can then simply automatically convert it into any of the remaining formats so that more people can benefit from it.

CBS API Changes Handling

When a Cloud-based Service changes its API, it will almost certainly bring significant changes on some of the existing Objects and/or the url schema itself, which will in turn bring a large transition overhead to the enterprises that utilize it, increasing development costs due to the required code refactoring and documentation updates. By utilizing the monitoring services of tools like API changelog [], changes on CBS APIs are captured and affected developers are notified so as to update the corresponding versions in the API Builder manually or by using one of the standards.

To the best of our knowledge, OPENi API Builder provides the easiest way to adapt all these changes through the swift reflexes of its community, shifting the API maintenance complexity from the enterprise to the Builder community of developers who are also interested in ensuring the sustainability of these APIs.

Then, all the Objects of an API created at the Builder can have their properties mapped to the corresponding properties of a similar Object of some CBS. These mappings can be thought of as arrays of labels, where each label represents a property from this particular CBS API. New versions can be drafted and linked to the appropriate CBS ones via a drag and drop mechanism of these labels. By connecting the Objects, a mashup is created that can afterwards be exported in all standards.

5   Conclusions

OPENi API Builder aims to leverage the power of the community to achieve sustainable APIs and seamless integration with multiple Cloud-based Services while having the ability to adapt its services to the business needs. Its carefully designed UI and all the developed functionalities around API management make API Builder a valuable platform for enterprises, start-ups and freelancers. Easy to use, clear and fully equipped with all the needed tools, Builder delivers documentation for its users' APIs in the most popular standards, providing compatibility with almost all business needs.

OPENi API Builder is available at Github as an open source project. As it is still a work in progress, all its provided functionalities have yet to be tested and validated by the developer community.

As future steps, dissemination activities will take place in order to extend the developer community of OPENi API Builder and establish its role as a useful API creation and management tool among enterprises. There are also plans for advancing the API Builder in a more automated tool for managing Cloud-based Services through different versioning. This process will include parsing an API documentation in one of the standards to automatically create the whole API in the Builder, hence accelerating the matching process. This has already been partially implemented for Swagger documentation to enable the interconnection of the Graph API Platform and the API Builder. All previously described major documentation standards can already be transformed from one to another, so what actually remains to be made is determining the versioning and deprecation policy required to guarantee the most comforting solution for developers, start-ups and enterprises when using the API Builder.

Acknowledgments. This work has been created in the context of the EU-funded project OPENi (Open-Source, Web-Based, Framework for Integrating Applications with Social Media Services and Personal Cloudlets), Contract No: FP7-ICT-317883.

References

1. Gat, I., Succi, G.: A Survey of the API Economy. Agile Product & Project Management. Executive Update **14**(6)
2. Chesbrough, H., Vanhaverbeke, W., West, J. (eds): Open innovation: Researching a new paradigm. Oxford University Press (2006)
3. Chesbrough, H., Bogers, M.: Explicating open innovation: clarifying an emerging paradigm for understanding innovation. New Frontiers in Open Innovation, 3–28. Oxford University Press, Oxford, Forthcoming (2014)
4. Programmable Web API Directory. http://www.programmableweb.com/apis (accessed: May 26, 2015)
5. Facebook versioning overview. https://developers.facebook.com/docs/apps/versions (accessed: May 26, 2015)
6. Dropbox API compatibility notes. https://www.dropbox.com/developers/core/docs (accessed: May 26, 2015)
7. Espinha, T., Zaidman, A., Gross, H.-G.: Web api growing pains: stories from client developers and their code. In: 2014 Software Evolution Week-IEEE Conference on Software Maintenance, Reengineering and Reverse Engineering (CSMR-WCRE). IEEE (2014)
8. Wang, S., Keivanloo, I., Zou, Y.: How do developers react to RESTful API evolution? In: Franch, X., Ghose, A.K., Lewis, G.A., Bhiri, S. (eds.) ICSOC 2014. LNCS, vol. 8831, pp. 245–259. Springer, Heidelberg (2014)
9. Robbes, R., Lungu, M., Röthlisberger, D.: How do developers react to api deprecation?: the case of a smalltalk ecosystem. In: Proceedings of the ACM SIGSOFT 20th International Symposium on the Foundations of Software Engineering. ACM (2012)
10. Dig, D., Johnson, R.: How do APIs evolve? A story of refactoring. Journal of Software Maintenance and Evolution: Research and Practice **18**(2), 83–107 (2006)
11. Alvertis, I., Petychakis, M., Lampathaki, F., Askounis, D., Kastrinogiannis, T.: A community-based, graph API framework to integrate and orchestrate cloud-based services. In: AICCSA 2014 (2014)
12. Petychakis, M., Alvertis, I., Biliri, E., Tsouroplis, R., Lampathaki, F., Askounis, D.: Enterprise collaboration framework for managing, advancing and unifying the functionality of multiple cloud-based services with the help of a graph API. In: Camarinha-Matos, L.M., Afsarmanesh, H. (eds.) Collaborative Systems for Smart Networked Environments. IFIP AICT, vol. 434, pp. 153–160. Springer, Heidelberg (2014)
13. Apiary based on API Blueprints. http://apiary.io/ (accessed: May 26, 2015)
14. RAML. http://raml.org/ (accessed: May 26, 2015)
15. Lanthaler, M., Gütl, C.: Hydra: a vocabulary for hypermedia-driven web APIs. In: LDOW 2013. APA (2013)
16. Hadley, M.J.: Web Application Description Language. W3C Member Submission (2009). http://www.w3.org/Submission/wadl/ (accessed: May 26, 2015)
17. Swagger: A simple, open standard for describing REST APIs with JSON. Reverb Technologies (2013). https://developers.helloreverb.com/swagger/ (accessed: May 26, 2015)
18. Hypertext Application Language. http://stateless.co/hal_specification.html (accessed: May 26, 2015)

19. JSON API Specification. http://jsonapi.org/ (accessed: May 26, 2015)
20. JSON-LD Specification. http://json-ld.org/ (accessed: May 26, 2015)
21. Collection+JSON Specification. http://amundsen.com/media-types/collection/ (accessed: May 26, 2015)
22. API Search Index. http://apis.io/ (accessed: May 26, 2015)
23. Mashape Application. https://www.mashape.com (accessed: May 26, 2015)
24. APIS.JSON Format for indexing into apis.io. http://apisjson.org/ (accessed: May 26, 2015)
25. Runscope Application for monitoring & testing APIs. https://www.runscope.com/ (accessed: May 26, 2015)
26. Apimatic application (create SDK). https://apimatic.io/ (accessed: May 26, 2015)
27. REST United (create SDK). http://restunited.com/ (accessed: May 26, 2015)
28. Apigility. https://apigility.org/ (accessed: May 26, 2015)
29. Strongloop Arc. http://strongloop.com/node-js/arc/ (accessed: May 26, 2015)
30. API Designer by Mulesoft. http://api-portal.anypoint.mulesoft.com/raml/api-designer (accessed: May 26, 2015)
31. Appcelerator Platform. http://www.appcelerator.com/platform/apis/ (accessed: May 26, 2015)
32. Apigee Edge API Platform. http://apigee.com/about/products/apigee-edge-and-apis/edge-api-services (accessed: May 26, 2015)
33. Kimono by kimonolabs. https://www.kimonolabs.com/ (accessed: May 26, 2015)
34. OPENi API Builder source code at Github. https://github.com/OPENi-ict/api-builder
35. Shah, S.K.: Community-based innovation and product development: findings from open source software and consumer sporting goods (2003)
36. Krishnamurthy, S.: An analysis of open source business models (2005)
37. Lee, G.K., Cole, R.E.: From a firm-based to a community-based model of knowledge creation: The case of the Linux kernel development. Organization Science **14**(6), 633–649 (2003)
38. API Changelog, documentation version monitoring. https://www.apichangelog.com/ (accessed: May 26, 2015)

End-User Centered Events Detection and Management in the Internet of Things

Stefano Valtolina$^{(\boxtimes)}$, Barbara Rita Barricelli, and Marco Mesiti

Department of Computer Science, Università degli Studi di Milano, Milano, Italy
{valtolin,barricelli,mesiti}@di.unimi.it

Abstract. With the widespread of Internet of Things' devices and sensors, the use of applications for combining data streams is increasing in a high number of dynamic contexts, with a huge variety of users having radically different backgrounds and a plethora of diverse tasks to perform. On the one hand, users need a system able to support them in detecting significant events with respect to their context of use. On the other hand, such users need an environment in which they can express proper requirements for enabling the system to react autonomously according to events in the real/physical world, by running processes that trigger actions and perform services. In this paper, we provide the design of a web environment that we are developing around the concept of event, that is simple or complex data streams gathered from physical and social sensors that are encapsulated with contextual information (space, time, thematic). Moreover, from a study of the most diffused applications for Internet of Things that provide the user with tools for controlling the combination of data streams coming from devices and sensors, we identify and discuss some open problems and proposes a new paradigm and language to address them and to be integrated in our web environment.

Keywords: Internet of things · Event detection · Pervasive computing · Unwitting developers · End users

1 Introduction

Nowadays we are witnesses of the proliferations of different sensor devices able to produce heterogeneous types of data (textual, visual, audio, and other rich multimedia formats) that can be profitable used for detecting, handling and advising people of the verification of particular conditions about physical phenomena (like temperature or humidity). These events are characterized according to different dimensions such as: the time (when they happen), the space (where they happen) and the thematic (what they concern) that can be exploited from one side for the identification of the useful information needed to face a given event and from another side for the analysis and forecast of useful activities to carry out. Conventional data analytics platforms cannot be exploited profitably for handling this kind of data and new advanced architectures should be developed for several reasons. In this context, network configuration, sensor detection and discovery are difficult issues to be solved. Moreover, sensors and

© Springer International Publishing Switzerland 2015
F. Daniel and O. Diaz (Eds.): ICWE 2015 Workshops, LNCS 9396, pp. 77–90, 2015.
DOI: 10.1007/978-3-319-24800-4_7

the data they produce should be handled in real time in order to be properly elaborated during the event. Therefore, scalable and efficient solutions should be devised that can be applied on-line. ETL (extract, transform and load) solutions, usually applied off-line, need to be revised and applied on-line for generating fresh and timely data. Finally, user-friendly environments have to be conceived for properly helping the users in combining the data coming from different sensors and devices and for enabling them to take under control the huge amount of data and events that can be generated. In this paper, we propose a novel event-based data model for manipulating multi-dimensional relations between multiple data streams based on event dimensions (spatial, temporal and thematic dimensions). Then, we propose a new language and paradigm for End-User Development that can be exploited in the context of Internet of Things (IoT). Specifically, we propose a sensor-based rule language able to support end users in aggregating and combining data originated by several sensors/devices. This language aims at enabling end user for unwittingly developing personalized IoT environments according to specific temporal, spatial, and fuzzy conditions that may affect the elements in the IoT environment.

The paper is organized as follows. In Section 2, the new architecture will be illustrated and compared with respect to conventional Data Warehouse (DW) systems. Section 3 describes the event data model tailored for handling multidimensional complex events according to a multigranular spatio-temporal-thematic (STT) approach. Section 4 presents the current state of the art of End-User Development (EUD) strategies in IoT and the applications that enable the users to arrange data coming from IoT devices/sensors. Section 5 starts with the ETL operations that are at the base of our event management model, and continues with the user interface developed for the application of such operations in a workflow based on the combinations of data coming from different sensors and devices. At the end of the section, the current state of the art of End-User Development (EUD) strategies in IoT is discussed along with the applications enabling the users to arrange data coming from IoT devices/sensors. Finally, Section 6 introduces a new EUD paradigm and language in IoT domain.

2   Research Background: From Conventional to Event based Analytic Platform

With OLAP ("Online Analytical Processing") systems, users can analyze data through the application of different cube operators applied to the data-warehouse (DW). For example, a company might want to compare their sales in June with sales in July, and then compare those results with the sales made in another location, which might be stored in a different database. Therefore, a data warehouse can be viewed as a large repository of historical data organized in a pertinent way in order to perform analysis and to extract interesting information using OLAP technology. Nevertheless, traditional data warehouse and OLAP systems are not able to support users in analyzing highly dynamic data processing and to generate meaningful analytical reports. This is particularly true in IoT systems where the data-streams are characterized by the 3V dimensions Volume, Variety, Velocity [1], along with the issue of integrating information coming from heterogeneous networks. In order to endow DWs and OLAP with these capabilities, recently the term "active" or "real-time" DW [2] was coined to capture the need for a DW containing data as fresh as possible. In traditional DWs, there is a distinction between ETL (Extract, Transform, and Load) and OLAP operations. The former are used for feeding the DW, whereas the latter are used to query the DW once the cube has been defined. Moreover, feeding the DW is considered a pre-processing phase that is taken offline with the purpose to solve many issues (extracting, cleaning and riconciliate heterogeneous data, and denormalize data) in order to make somehow simpler the execution of the OLAP operations. Stream DW have been also proposed for handling information produced in streams and approaches based on sliding windows are used to manage continuous data [3].

In IoT the periodic population of a DW is considered outdated and new ETL and OLAP services are required in order to work in real time (or almost real time), that is, data are loaded in real time from the OLTP (Online Transaction Processing) systems into the DW, providing a convenient way for user to real-timely read the data information and make tactical decisions. In this context, the ETL and OLAP operations must be tightly connected for making the process possible and their implementations need to be quite efficient and scalable.

In our work, we wish to develop real-time DW that is developed around the ECA (Event, Condition, and Action) paradigm [4] because it is the most promising approach for the development of the behavior of an intelligent environment. We identify an event with both its temporal, spatial and thematic dimensions that can be exploited from one side for the identification of the useful information needed to face a given event and from the other side for the analysis and forecast of useful activities for notifying people. First, we need to point out the adoption of a different data model. Indeed, conventional DWs adopt the relational model for representing stored information, but this is not adequate in Event-based warehouse because of the heterogeneity of the data formats and also the fact that information is stored at different granularities (ranging from simple raw values to structured and complex data). Second, conventional systems process stored data, whereas in the Event-based warehouse also streams of data should be processed, collected and analyzed through temporal windows. Contextual information can be associated with the stream by means of machine learning algorithms and exploited for extracting knowledge. For what concerns the implementation of the architecture and of the required operations, we are considering the use of cloud-based architectures because we need to guarantee high throughput and low end-to-end latency from the system, despite possible fluctuations in the workload. We are currently evaluating several architectures like Apache S4 [5], D-Streams [6], Storm [7] and StreamCloud [8].

3   STT Event Data Model

In order to represent events, we consider a set of basic and structured types (like list and records) whose values are represented thorough a JSON-like notation. These types allow the representation of different aggregation of values without imposing the restrictions of the relational model and so are able to handle a great variety of datatypes. Moreover, we consider spatial and temporal types at different granularities. Temporal granularities include seconds, minutes, days with the usual meaning adopted in the Gregorian calendar, whereas, meters, kilometers, feet, yards, provinces and countries are examples of spatial granularities. Different granularities provide different partitions of their domains because of the diverse relationships that can exist among granularities, depending on the inclusion and the overlapping of granules [9].

For example, in temporal granularity seconds *is finer than* minutes, and granularity months *is finer than* years. Likewise, spatial granularity provinces *is finer than* regions. Thematics are also considered for associating a semantic annotation to a given value. For example, the annotation "very high" can be associated to the event representing the degree (high) of temperature that is detected in a given area. Several thematics can be identified depending on the context where the datum is acquired or processed. Thematic can indeed be inferred by machine learning algorithms. Relying on the concepts of temporal and spatial granularities, we present the concept of event, that is a value associated with a spatial object at a given time according to given thematics. Therefore, an event is a value represented at a given spatio-temporal granularity for which thematic information is added. Relying on the concept of event, we can characterize an event stream that a source can produce. A source can be sensor (either physical or virtual) or a service (for example by aggregating sensors/services streams). As an example, let us consider a gym scenario where a trainer wishes to monitor the physical conditions of her/his users in order to suggest better exercises. These exercises need to be compliant to user's biological parameters such as the heartbeat variability or the weight and they need to be used on a fitness device such as a treadmill according to a plan of execution (the time that the user has to run, speed and inclination of the treadmill). The simple events, collected by these types of sensors and devices, can be characterized by the time dimension (when the biological parameters have been detected), the space dimension (where the biological parameters have been detected: at home, into the gym, or where the treadmill is located). The thematic of a simple event is the meaning associated to a value or a set of values. For example, the event: "Tachycardia" can be related to a very high number of heartbeats. Instead, the event: "Hard run" can be related to high inclination of the treadmill that works fast. The outcome the trainer expects is a set of complex events that are the actual exercises to provide to the users according to their physical conditions. In addition, these events have time dimension (when the exercise has to be executed) and space dimension (where the exercise has to execute in case the gym has more than one treadmill of different types). Moreover they have a thematic, which in this case is a complex document representing the execution plan of the exercise and the related user's physical condition (as a combination of heartbeat variability and weight).

4   End-User Development and Internet of Things

Shneiderman defined in [10] his concept of "old computing", focused on what computers can do for the user. On the other hand, today we deal with the "new computing" concept that refers to people activity and what people can actively do by using computers. Users of digital devices and interactive systems are increasingly evolving from passive consumers of data and computer tools into active producers of information and software [11][12]. The potentials offered by network and connectibility of the objects do not only enrich the person's personal sphere but also give the chance of sharing data with other people like family members, friends, colleagues, or others. In this scenario, end users find themselves at the center of a complex scenario that they need to manage in efficient, effective, and satisfactory way. EUD represents the ideal approach for empowering end users and make them becoming unwitting developers in their own IoT environment [13][14][15]. As widely reported in the literature, EUD can be enabled by applying methods and techniques and by offering specific tools that allow end users to develop solutions without having specific programming skills and knowledge about programming languages. Specifically, the solutions offered by EUD include tools for the customization of applications by parameters setting, control of a complex device (like a home-based heating system), and even scripting of interactive Web sites [16]. EUD allows users to configure, adapt, and evolve their software by themselves [17] and such tailoring activities, together with personalization, extension, and customization are defined in literature in different ways, sometimes referring the same concepts and sometimes referring different ones [18]. The target of EUD activities in the IoT context is not (only) the user interface and the behavior of an interactive system, but embraces the whole IoT scenario of sensors and devices. Therefore, we need to distinguish between those activities that can be made at three different levels: hardware, software, and data. EUD activities on hardware are those made on the devices via their bundled applications. They typically are configuration, personalization, and customization by setting parameters and choosing among existing behaviors. The activities on software target concern the applications used for controlling more than one sensor/device (even of different brands) and include tailoring by integration of existing and/or new functionalities, macros, visual programming, and programming by examples. The EUD activities that can be made on data can be resumed in aggregation, filtering, and porting. In what follows, we will use the classifications presented so far for discussing the state of the art of applications that can manage data originated by more than one sensor/device and shared on Social Media, and that enable end users to unwittingly develop their own IoT environment. In Section 6, we provide an overview of the current state of the art in IoT applications/tools that offer EUD activities to their users and we relate them with the new paradigm that we propose in this paper.

5   Event Management

In our STT data model, one of the most important EUD activity takes place during the design of the most relevant and significant events that characterize a context of use. For example in a gym, the trainers need to combine data coming from different sensors and devices for properly defining the exercises. These EUD activities are exploited for executing efficiently and effectively Event ETL operators over a programmable network (e.g., filter by in-network data processing). So far, we have developed three kinds of operators that are relevant in our context: conversion, merging, and connecting operators. Conversion operators are used for changing the spatio-temporal-thematics granularity of an event stream. Merging operators are used for making the union of events at the same spatio-temporal granularity in a single data structure. Connecting operators are used to link a data source to another one by applying a SQL joint-like operator. In the rest of the section, we focus on the graphical visual interface using which the domain expert, not necessarily expert in IoT technologies, can identify the sensors and apply the operators for converting, merging and combining their data stream.

5.1   Visual Event ETL

ETL operations have been proposed in different contexts depending on the kinds of data to handle (structured and semi-structured). In [2][3][19] there is a good treatment at the conceptual level for feeding a DW. Moreover, in [20] there are approaches for the semi-automatic generation of ETL operations depending on the user needs and context of use. ETL operations are usually coupled with graphical visual data-flow for helping the user in the identification of the original data sources, the application of the operations for extracting, cleaning, transforming and combining their data. In our case, the user is a domain expert that needs a visual interface for combining sensors, and services that are used for generating a flow of significant events to be provided to the final user. For example in a gym, a trainer possibly needs to integrate fitness devices with personal data of her/his users and with other data streams coming from personal sensors used for capturing biological parameters. Once the ETL specification is completed, some strategies are proposed for the optimization of the dataflow and for the efficient execution of the loading schedule. In the fitness case study, the events might concern the user's physical condition, which will be used for suggesting what machine is better to use or which exercises. These approaches have been mainly developed for producing relational data to feed conventional DW system. In [21] an approach is presented for feeding arbitrary target sources (either relational or based on a NoSQL system). In our context, we designed a Web environment where the domain expert can drag and drop different sensor data sources and visually apply on them a set of operations. This application offers an engine and graphical environment for data transformation and mashup. As depicted in Fig. 1 (section A), this mash-up consists of a user interface that contextually displays icons of data sources or operations in order to link, filter or merge data coming from different sensors. It relies on the idea of providing a visual workflow generator for letting the domain experts to create aggregation, filtering, and porting of data originated by sources. An advanced use of such visual paradigm allows the domain experts to have an online generation of sample data coming from the data sources dragged-and-dropped on the canvas, or as result of the operations carried out on them (see Fig. 1 section B). By this strategy, information is gathered from the net and specific filters and operations can be triggered on user requests. This is a very simple and easy to learn solution based on the definition of sources to use for collecting data and on the possibility to apply our operations in a visual way. Future activities aim at developing machine-learning algorithms for exploiting the possible operations that domain experts can apply to a set of data sources or for providing them useful predictions of what can happen by integrating the selected data.

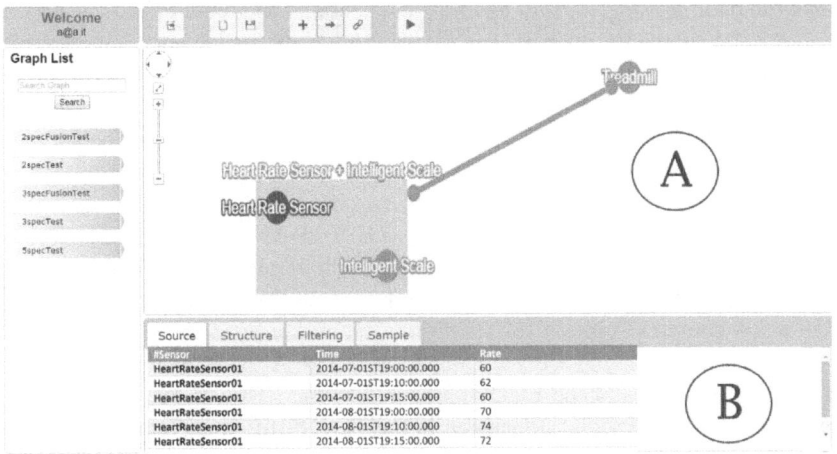

Fig. 1. A screenshot of the user interface of the Visual Event ETL

6   End-User Empowerment for Controlling Events

Currently several tools in IoT context are applying the ECA paradigm for supporting users in generating personalized actions according to a set of self-defined conditions. Users can monitor through them specific events of their environment, and some examples are IFTTT (https://ifttt.com) and Atooma (http://www.atooma.com). By analyzing these services, we identified two main types of applications that differ in terms of activities and interaction style. The first one allows users to define sets of desired behaviors in response to specific events. This is made mainly through rules definition-wizards that rely on the sensors/devices states. Rules can be typically chosen among existing ones or can be tweaked through customization. These EUD activities put in place a task automation layer across all sensors/devices in the IoT environment. Such strategy is adopted by those applications that use automated rules-based engines like Atooma and IFTTT– by using the programming statement IF this DO that, and by Wewiredweb (https://wewiredweb.com/) with the statement WHEN trigger THEN action. A more advanced use of these applications sees the involvement of recommendation systems (RSs) as part of the IoT ecosystem. In these cases, EUD activities are supported by the RSs that, relying on end user's pattern of use of devices, can suggest compelling examples of statements that the user can adapt to his needs.

A second type of applications stems from the outstanding work done with Yahoo's Pipes (https://pipes.yahoo.com/pipes/). These solutions typically use formula languages and/or visual programming. Applications like Bipio (https://bip.io/) and DERI pipes (http://pipes.deri.org/) offer engine and graphical environment for data transformation and mashup. They are based on the idea of providing a visual pipeline generator for supporting the end user in creating aggregation, filtering, and porting of data originated by sources. An advanced use of such visual paradigm is offered by WebHooks (https://developer.github.com/webhooks/) that allows the end users to even write their personal API for enabling connections with new sources of data. Both propose EUD strategies, adoptable in the context of the IoT applications, for gathering information from across the net and triggering specific actions when certain things happen. The first type of applications offers a very simple and easy to learn solution based on the definition of ad hoc rules that can notify the end users when something happens – e.g. when their favorite sites are updated, when they check-in in some places or their friends do, or warn them when specific weather conditions are going to take place. However, the adoption of the IF-THIS-THAN-THAT/WHEN-TRIGGER-THEN-ACTION patterns is not enough to deal with more sophisticated rules based on time, space, and fuzzy conditions. On the other hand, the second type of applications offers a too complex solution for supporting the end user in expressing their preferences. Pretending that end users are able to deal with APIs of several sensors/devices put at risk the success of the EUD approach. Another issue of the current proposals regards the use of the social dimension in the most diffused applications, while time and space dimensions are almost neglected.

To face these problems, in the next section we propose an extension of the IF-THIS-THEN-THAT paradigm by presenting a sensor-based rule language able to support the end user in defining rules in a more articulated way but keeping the complexity at an acceptable and accessible level. This idea is to keep the simplicity of the IF-THIS-THEN-THAT paradigm pairing it with the use of formula languages. Moreover, time and space dimension will be exploited and fuzzy conditions adopted for expressing more loose rules in the statements.

6.1   A New EUD Paradigm and Language for IoT

In order to manage the great flow of events generated by the event-based DW systems and the flow of data-streams provided by sensors it is necessary to devise an innovative strategy for supporting the end user empowerment. To make the users feel like they are in control, an ECA-based rule strategy can be a promising solution. Through the definition of a set of conditions, users can enable the system to suggest some suitable actions according to their needs and context of use. The flow control determines how the system will respond when certain conditions are in place and specific parameters are set. By apply this strategy, the end user needs to state a if condition (e.g. the weather station says that it is going to rain in the next 12 hours) and an action (e.g. tweet this news on the end user's Twitter personal account using hashtags #weather #rain). As described before, this solution is adopted in many applications for supporting end users in creating rules for their IoT environment. The IF-THIS-THEN-THAT paradigm seems to work well when end users need to be warned or notified on a specific event, but uses a very simple language that has a quite low expressive power. By exploiting the three dimensions defined for our STT Event Data Model, we propose a new rule-base language able to give to end users the possibility of setting triggers that do not depend just on one simple event but also on complex events. The introduction of the time dimension allows end users to set triggers that can be fired at some specific time, delayed in case of certain conditions are verified, and may be repeated until some event happens. The space dimension gives end users the chance of linking triggers to the place/area where they currently are, where they will possibly be in the future, where they are moving into, or where some events are taking place. In [4][22] there is a nearly unanimous agreement on an extension of "classical" trigger languages by including time dimension.

The proposal in this field can be summarized as follows: (1) Rules should be triggered by the occurrence of the dimension of events, (2) Enabling periodically repeated triggering of the same reaction, where the period is specified by an expression returning a time duration. (3) Delaying reaction execution to some later point in time relative to the triggering event of a rule. In [23] the authors propose a set of functionalities to be implemented with triggers written in SQL:1999 standard that cover three types of temporal categories – absolute, periodic, and relative event specifications – and allow to base delay or periodic repetition on valid time or transaction time events, respectively. According to this proposal of functionalities, we can provide users with a new set of rules composition-strategies able to go beyond the simple use of an IF-THIS-THEN-THAT statement. Up to now, rules can be triggered by call events only, and reactions are always executed one time. We identified four types of rule:

1. Space Events: rules that need to be triggered if the data stream refers to a specific geographical place/area. An example: IN *"home place"* IF *"I do not enter in the gym-room at least three times in a week, send me an alert"*.
2. Time Events: Rules triggered on certain absolute time events are the most common feature of time-triggers. An example: AT *"summer time"*, IF *"my sleep-controller detects that I did not sleep well last night"* THEN *"my activity tracker device should suggest me to take a walk before going to bed"*.
3. Delayed Reaction Execution: Reaction execution can be delayed by combining a call event with a temporal offset. This offset is a time-valued attribute of the related environment, thus generating "relative events". For instance, to check three months after my last blood test if I need another test, a possible rule could be: AT *"The date of my last blood test + 3 months"* IF *"the person scale says that I lost more than 10 kilograms"* THEN *"my smart watch should show a message suggesting to book a medical exam"*.
4. Repeated Reaction Execution: Repeating execution of a particular action regularly after a fixed period has passed. In this case the keyword EVERY could be combined with an expression of type PERIOD, e.g. EVERY MONTH, or EVERY 3 HOURS, or with even more sophisticated specifications, such as EVERY MONDAY, EVERY 2nd MONDAY IN A MONTH, or EVERYDAY EXCEPT SATURDAY.

The EUD paradigm we propose in this paper aims supporting the end user in composing such space/time-based rules for extending the well-established but not powerful IF-THIS-THEN-THAT paradigm. Our Sensor-based Rule Language follows syntax and semantics of a Policy Rule Language was proposed in [25], and is based on the ECA paradigm. Our language allows the specification of rules stating policies for triggering actions (one or a set). The general format of a rule is the following (square brackets denote optional components):

```
RuleName: "MY RULE"
ON SOURCE[s]
  [WHENEVER "Condition"]
Action: "Some Actions"
  [VALIDITY: Validity_Place-Interval]
```

A rule consists of several components. The `RuleName` component represents the rule identifier. Users can retrieve rules by means of such identifier for visualizing, sharing, dropping, or modifying them. `SOURCE[s]` represents the source or set of source upon which data the rule is triggered. In our case the user can take data from a sensor for example a thermometer, a service for example the weather forecast service or an event-based DW system. The last source provides users with the possibility to take into account complex events generated by domain experts in specific context of use as for example in the fitness case study previously described or in other contexts such as traffic control or healthcare or emergency situations.

Each source exposes a set of data or events, which can be used for expressing the conditions. `Condition` is an optional conditional expression. `Action` is an expression that states what happens when the condition is verified. `Validity_Place-Interval` is a special spatial and/or temporal condition also expressed by means of the condition language we developed, representing the space and time period during which the rule is enabled. For example, if the interval [EVERYDAY EXCEPT SATURDAY] is specified we know that a rule is enabled every day of the week but not on Saturday. However, if `Validity_Place-Interval` is not specified, we know that the rule is always enabled. By means of `Validity_Place-Interval` it is possible to state that certain rules are not always enabled; rather, they are enabled only if an event happens in a specific place or during specific temporal intervals. Such a feature is not provided by conventional apps for IoT. A possible rule generated in the fitness case study might be:

```
RuleName: "Next week exercises"
ON GYM Event-Base DW AND Thermometer AND Weather forecast
service
  WHENEVER "the suggested exercises concern the use of a
treadmill" AND "the temperature is not more than 30 de-
gree" AND "the weather is forecasted good"
Action: "suggest me to run outside into the park"
  VALIDITY: IN "Milan" and AT "June-August"
```

In case the user involves event-based DW systems in the definition of the rule, she/he has the possibility to trigger more complex actions according to the parameters of the events provided by the DW. For example if the trainer describes the exercise with an indication of the rate of difficulty the user can create a new rule for asking for a simpler exercise in case the sleep controller detects that last night she/he slept few hours.

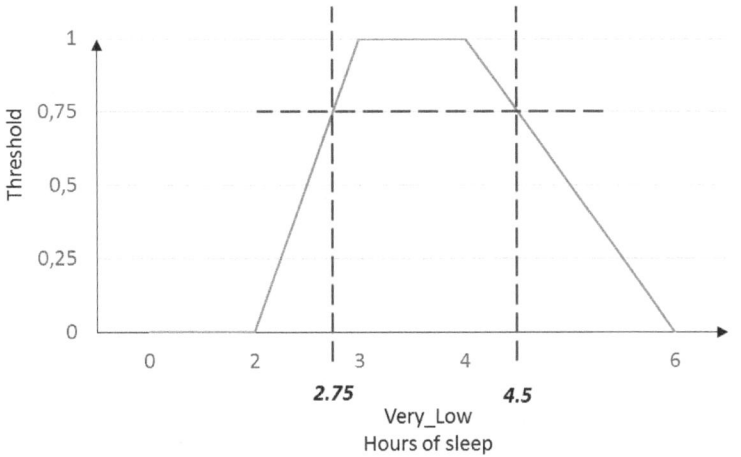

Fig. 2. Example to illustrate hours of sleep distribution.

Another extension of the **IF-THIS-THEN-THAT** paradigm is to provide end users with a more flexible way for expressing the condition statement by incorporating fuzziness into the condition on which the events need to be triggered. This extension focuses on the linguistic values of the fuzzy condition [25], in which the fuzziness concerning the linguistic concepts is interpreted in an application context. A linguistic variable is represented by a quintuple of <v,T,X,g,m> where v is name of the linguistic variable, T is set of linguistic terms applicable to variable v, X is the universal set of values, g is the grammar for generating the linguistic term, m is the semantic rule that assigns to each term t∈T, a fuzzy set on X. To illustrate our approach, we use as example a sleep monitor. Let represent a linguistic variable with a graphical distribution based on four parameters as depicted in Figure 3. Very_Low for hours of sleep is represented using trapezoidal function as Very_Low (2 hours, 3 hours, 4 hours, and 6 hours). Current IoT applications use simple statements such as "Hours of sleep <= 3 hours" to indicate when the value is very low. Using our language, users can use the statement "Quantity = $Very_Low": a set of values are related to "$Very_Low" in this comparison, rather than one single value. Fulfillment threshold is allowed to specify the condition with a degree value in the range of [0, 1]. For example, in Figure 3, we used 0.75 as the threshold to indicate that the value of hours of sleep is very low with the degree of 0.75. As a result, a value in the range [2.75, 4.5] indicates that "the number of hours of sleep is very high with a threshold of 0.75". By using our Rule Language the user can express a fuzzy condition is this way:

```
RuleName: "Quality of sleep Monitor"
ON sleep-controller AND Thermometer
   WHENEVER "the number of hours of sleep is $Very_Low"
AND "the temperature is not $Very_High"
Action: "The activity tracking device suggests me to take
a walk before going to sleep"
```

7   Conclusion and Future Steps

In the paper, we presented the architecture of a web environment for gathering, computing, and diffusing data originated and streamed by physical and social sensors. The architecture is based on event data model designed for an approach that focuses on space, time and themes. After presenting strategies of End-User Development in the IoT context, aimed at giving the end users more freedom and power to assemble different data sources/sensors in an ad-hoc and personalized solution, we proposed a EUD paradigm and related language that is currently under development in an IoT application for wellness domain.   In order to manage the overload of information coming from different sensors and event-based DW systems and to mitigate the overflow of activities that the user has to perform for taking all under control, future studies aim at investigating a solution for predicting ECA rules according to the user's behavior and practices. In a unware way the user will have the possibility to experiment new rules predicted by an intelligent system by applying a serendipity-based strategy. The suggestion can be formulated by taking into account the user's profile and recurring activities carried out by using the combination of sensors and event-based DW systems. Moreover, a rating process of the suggestions provided to the user will enable an effective human control over the intelligent algorithm for improving the set of recommendations. This because by relying only on automatic suggestions may not be entirely appreciated by the user, which risks to get frustrated in using the recommendation system whenever the recommendations prove to be wrong for her/him. To solve this problem, the main challenges that we have to face are: (i) The need of a recommendation model and, as a consequence, of a recommendation policy understandable by the user; (ii) The need to define how the user can be part of the recommendation process; (iii) The need to exploit the social relationships between users. User-centered recommendations can be enriched in considering social networks where each user will certainly appreciate receiving recommendations from those considered "closest" to her/him; (iv) The need to take into account the uncertainty of both the ratings/suggestions proposed by the social networks members. In fact, judgments collected from a plethora of users with different habits and cultures may produce contradictions, which in turn result in rules having a high degree of ambiguity. This calls for a granular interpretation of the information provided by such crisp attributes, for instance in terms of fuzzy sets, rough and interval sets, and so on.

In conclusion, the definition of an innovative system able to balance the human and the machine empowerment by adopting a programming by practices paradigm will allow us the unprecedented possibility to assist users or groups of users in completing their daily tasks in an unaware way.

References

1. Schroeck, M., Shockley, R., Smart, J., Romero-Morales, D., Tufano, P.: Analytics: The Real-World Use of Big Data (IBM Institute for Business Value - Executive Report). IBM Institute for Business Value (2012)
2. Gorawski, M., Gorawska, A.: Research on the stream ETL process. In: Kozielski, S., Mrozek, D., Kasprowski, P., Małysiak-Mrozek, B. (eds.) BDAS 2014. CCIS, vol. 424, pp. 61–71. Springer, Heidelberg (2014)
3. Imran, M., Castillo, C., Diaz, F., Vieweg, S.: Processing Social Media Messages in Mass Emergency: A Survey eprint arXiv:1407.7071 (2014)
4. Widom, J., Ceri, S.: Active Database Systems. Morgan Kaufmann Publisher (1996)
5. Zhou, H., Yang, D., Xu, Y.: An ETL strategy for real-time data warehouse. In: Wang, Y., Li, T. (eds.) Practical Applications of Intelligent Systems. AISC, vol. 124, pp. 329–336. Springer, Heidelberg (2011)
6. Neumeyer, L., Robbins, B., Nair, A., Kesari, A.: S4: distributed stream computing platform. In: IEEE Int'l Conf. on Data Mining Workshops, pp. 170–177 (2010)
7. Zaharia, M., Das, T., Li, H., Hunter, T., Shenker, S., Stoica, I.: Discretized streams: fault-tolerant streaming computation at scale. In: ACM Symposium on Operating Systems Principles, pp. 423–438 (2013)
8. Marz, N.: Storm: Distributed and fault-tolerant realtime computation (2012)
9. Gulisano, V., Jiménez-Peris, R., Patino-Martinez, M., Soriente, C., Valduriez, P.: Streamcloud: An elastic and scalable data streaming system. IEEE Trans. Parallel Distrib. Syst. **23**(12), 2351–2365 (2012)
10. Shneiderman, B.: Leonardo's Laptop: Human Needs and the New Computing Technologies. MIT Press (2002)
11. Fischer, G.: Beyond 'Couch Potatoes': From Consumers to Designers and Active Contributors" (2002). http://firstmonday.org/issues/issue7_12/fischer/ (accessed on January 9, 2015)
12. Costabile, M.F., Fogli, D., Mussio, P., Piccinno, A.: Visual Interactive Systems for End-User Development: a Model-based Design Methodology. IEEE TSMCA **37**(6), 1029–1046 (2007)
13. Costabile, M.F., Mussio, P., Parasiliti Provenza, L., Piccinno, A.: End users as unwitting software developers. In: Proc. of WEUSE 2008, pp. 6–10. ACM (2008)
14. Costabile, M.F., Mussio, P., Parasiliti Provenza, L., Piccinno, A.: Advanced Visual Systems Supporting Unwitting EUD. In: Proc. of AVI 2008, pp. 313–316. ACM (2008)
15. Barricelli, B.R., Marcante, A., Mussio, P., Parasiliti Provenza, L., Valtolina, S., Fresta. G.: BANCO: a Web Architecture Supporting Unwitting End-User Development. IxD&A, 5-6, pp. 23–30 (2009)
16. Sutcliffe, A., Mehandjiev, N.: Introduction of Special Issue on End-User Development. CACM 47(9), 31–32 (2004)
17. Pipek, V., Rosson, M.B., de Ruyter, B., Wulf, V.: Introduction. In: Proc. of IS-EUD 2009, pp. V–VI. Springer (2009)
18. Mørch, A.: Three levels of end-user tailoring: customization, integration, and extension. In: Computers and Design in Context, pp. 51–76. MIT Press (1997)
19. Camossi, E., Bertino, E., Mesiti, M., Guerrini, G.: Handling expiration of multigranular temporal objects. J. Log. Comput. **14**(1), 23–50 (2004)
20. Vassiliadis, P., Simitsis, A., Skiadopoulos, S.: Conceptual modeling for ETL processes. In: Proc. Int'l Workshop on Data Warehousing and OLAP, pp. 14–21 (2002)

21. Mesiti, M., Valtolina, S.: Towards a user-friendly loading system for the analysis of big data in the internet of things. In: IEEE Computer Software and Applications Conference - Workshops, pp. 312–317 (2014)
22. Ceri, S., Cochrane, R., Widom, J.: Practical applications of triggers and constraints: Success and lingering issues. In: Proc. of VLDB 2000, pp. 254–262 (2000)
23. Behrend, A., Dorau, C., Manthey, R.: SQL triggers reacting on time events: an extension proposal. In: Grundspenkis, J., Morzy, T., Vossen, G. (eds.) ADBIS 2009. LNCS, vol. 5739, pp. 179–193. Springer, Heidelberg (2009)
24. Bertino, E., Cochinwala, M., Mesiti, M.: UCS-Router: a policy engine for enforcing message routing rules in a universal communication system. In: Proc. of Mobile Data Management 2002, pp. 8–16 (2002)
25. Jin, Y., Bhavsar, T.: Incorporating fuzziness into timer-triggers for temporal event handling. In: Proc. of IRI 2008, pp. 325–329 (2008)

Proposals for Modular Asynchronous Web Programming: Issues and Challenges

Hiroaki Fukuda[1]([✉]) and Paul Leger[2]

[1] Information Science and Engineering, Shibaura Institute of Technology,
Tokyo, Japan
hiroaki@shibaura-it.ac.jp
[2] Escuela de Ciencias Empresariales, Universidad Católica Del Norte,
Región de Antofagasta, Chile

Abstract. Because of the success in the Internet technologies, traditional applications such as drawing and spreadsheet software are now provided as web applications. These modern web applications adopt asynchronous programming that provides high responsive user interactions even if an application works without multi-threading. At the same time, as the scale of these applications becomes large, modular programming becomes important because it allows developers to separate concerns, meaning that the evolution of one module does not affect other modules. However, applying asynchronous and modular programming is difficult because asynchronous programming requires uncoupling of a module into two sub-modules, which are non-intuitively connected by a callback method. The separation of the module spurs the birth of other two issues: *callback spaghetti* and *callback hell*. Some proposals have been proposed without the lack of issues about modular programming. In this paper, we compare and evaluate these proposals applying them to a non-trivial open source application development. We then present a discussion on our experience in implementing the application using these proposals. Finally, we point out challenges that this kind of proposal should overcome toward a modular programming.

Keywords: Virtual block · Asynchronous programming · Aspect-oriented programming

1 Introduction

With the growth of high speed network and the variety kinds of computers that possess high computation capabilities, traditional standalone applications such as drawing and spreadsheet software are now provided using web technologies, namely web applications. Compared to traditional web applications, such modern applications adopt asynchronous behavior such as ajax that provides high responsive user interaction even if an application works without multi-threading. At the same time, the scale of such applications becomes large where modular programming becomes important because it allows separating concerns to be localized, meaning that the modification of one concern does not affect other concerns (*e.g.,* other modules). The basic idea of asynchronous programming is to decompose a blocking

© Springer International Publishing Switzerland 2015
F. Daniel and O. Diaz (Eds.): ICWE 2015 Workshops, LNCS 9396, pp. 91–102, 2015.
DOI: 10.1007/978-3-319-24800-4_8

operation that waits for its completion into a non-blocking operation that immediately returns control. Whenever the non-blocking operation execution ends, a piece of code known as *callback method*, is invoked. Therefore, the use of callbacks in asynchronous programming comes with issues that affect the modular development of software. The most notable two issues are *callback spaghetti* [7] and *callback hell* [8]. Callback spaghetti refers to the concern implementation that has a complex and tangled control structure because of continuations over callback methods. Callback hell refers to deeply-nested callbacks that have dependencies on data returned from previous asynchronous invocations. Some proposals have used to address these issues such as *async/await* from C# [1], *Promise pattern* [3] from JavaScript and SyncAS [4] from ActionScript. This paper compares and evaluates these proposals applying them to a non-trivial open source application development, called FlickrSphere [10] that is originally implemented by ActionScript3 and uses nested iterative callbacks, leading complicated control flows. The paper then presents a discussion on our experiences in its implementation.

Paper Roadmap. Section 2 gives an introduction to asynchronous programming and its problems. Section 3 briefly describes about FlickrSphere. In Section 4, we apply each one of three proposals to FlickrSphere to address asynchronous issues, we then present our experiences. Section 5 presents conclusions and future work.

2   Asynchronous Programming Problems

Asynchronous programming is now widely-adopted between mainstream programmers [1]. This section briefly describes asynchronous programming comparing to synchronous programming.

2.1   Synchronous Programming

Synchronous programming is the standard style used by programmers to write pieces of code. Listing 1.1 shows two classes: ImageViewer and Request. The ImageViewer class contains two methods: showFromURL and checkAndConvertToImage. The first method downloads data from a url and uses the second method to convert this data to an image if it passes an integrity check. The Request class has the send method that actually downloads some data by using the download method of the Downloader class. For this example, we assume download is a blocking operation, which takes a significant period of time. Because of several advantages such as reusability, maintainability and adaptability, dividing a system into a composition of modules is natural [9]. Therefore, in Listing 1.1, Request encapsulates how to get data from sources. As shown in Listing 1.1, in synchronous programming, we can obtain the result of a method invocation directly, then pass it to the next invocation as an argument, making the control flow clear.

2.2   Asynchronous Programming

Asynchronous programmign style makes it complicated to understand the control flow from pieces of code. Listing 1.2 shows the rewritten program of

Listing 1.1 replacing a blocking operation (download) with an non-blocking operation (downloadAsync). Three major changes are found in Listing 1.2. First, invoking checkAndConvertToImage is removed from showFromURL because send, that invokes downloadAsync, returns immediately without any data. Instead, the reference of the showFromURL is passed as a callback by using next, which is defined in Request. Second, a variable checksum must be defined in ImageViewer to keep the argument checksum in showFromURL because checkAndConvertToImage only accepts one parameter. Third, the show function call must be moved to checkAndConvertToImage because showFromURL does not contain the image. As described previously in this section, a module that uses a non-blocking operation requires representing the continuation as a callback. As a consequence, if the location of the continuation is far from the call-site, understanding control flow is difficult, leading to callback spaghetti. Moreover, a change of implementation inside of one module (*e.g.,* Request) may require call-site modifications in other modules (*e.g.,* ImageViewer), breaking modular principles. Besides, understanding control flow becomes difficult at a glance.

```
class ImageViewer {
  function showFromURL( url :URL, checksum : Function ): void {
    var e : Event = new Request().send( url );
    var img : Image = checkAndConvertToImage( e , checksum );
    show( img );
  }
  function checkAndConvertToImage( e : Event , checksum : Function ) {
    if ( checksum( e . data ))
        return convertToImage( e . data );
    else //throw an exception
} }
class Request {
  function send( url :URL): Event {
    return new Downloader().download( url );
} }
```

Listing 1.1. A synchronous version of a remote image viewer.

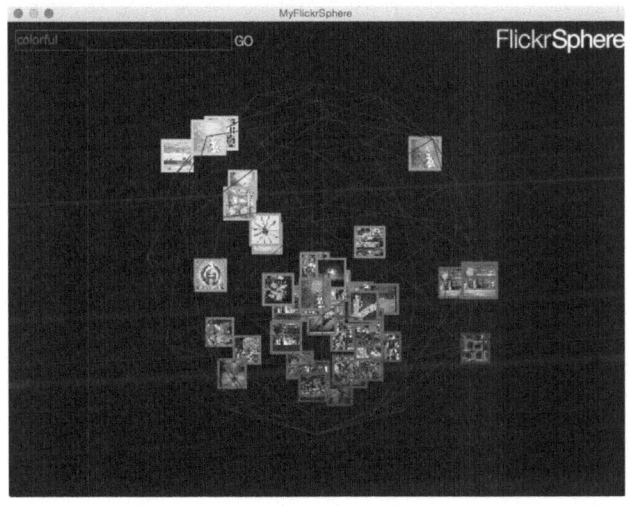

Fig. 1. A screenshot of FlickrSphere.

```
class ImageViewer {
  var checksum: Function;

  function showFromURL( url :URL, checksum : Function ): void {
    this.checksum = checksum;
    var request = new Request().next(checkAndConvertToImage);
    request.send( url );
  }
  function checkAndconvertToImage(e: Event ): Image {
    if (this.checksum(e.data)) {
      var img: Image = convertToImage(e.data );
      show(img);
    }
    else //throw an exception
} }

class Request {
  var nextF: Function;
  function next( f : Function ): void {
    nextF = f;
  }
  function send( url :URL): void {
    var dl = new Downloader();
    dl.addEventListener(Downloader.Complete, callback );
    dl.downloadAsync( url );
  }
  function callback(e: Event ): void {
    nextF(e);
} }
```

Listing 1.2. An asynchronous version of a remote image viewer.

3   Nested and Iterative Asynchronous Executions

FlickrSphere is an open source Web application implemented in ActionScript3.
Since ActionScript3 runtime does not provide *threads* for concurrent executions,
programmers need to use asynchronous programming if needed. This section
briefly describes the behavior of FlickrSphere that uses nested and iterative
asynchronous executions, then explains its original implementation.

3.1   FlickrSphere in a NutShell

FlickrSphere accepts keywords from users, and accesses to the flickr web service [2] to get all URLs matched by the keywords. Then FlickrSphere downloads
all images according to these URLs. Everytime a image is completly downloaded,
FlickrSphere displays the image on a circle that spins in the center of the screen.
If a user searches during downloading images, FlickrSphere cancels current downloads, then starts a new search after the completion of the cancel operation.
In addition to these main behaviors, FlickrSphere provides *Demo Mode* that
plays a search with a certain keyword automatically to show the behavior of
FlickrSphere.

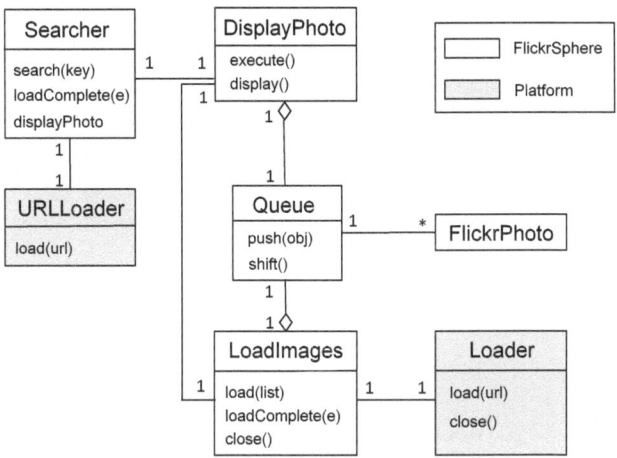

Fig. 2. A class diagram of the original implementation in FlickrSphere.

3.2   FlickrSphere Implementations

FlickrSphere is an open source application that shows images downloaded from
Flickr [2]. Figure 1 and Figure 2 show a screenshot and a simplified class diagram
of the original FlickrSphere[1] respectively. The main behavior of FlickrSphere is
provided by two phases: downloading a list of URLs matched by a given keyword,
and displaying each image after the download is completed. The main behavior
carries out nested and iterative asynchronous executions. Listing 1.3 shows the
piece of code that executes these two phases. For the first phase, **Searcher** directly
uses **URLLoader** that is provided by the Flash runtime and uses a non-blocking
operation (*i.e.*, **load** in Line 5), meaning that a callback (*i.e.*, **loadComplete** in
Line 7) is required. The **loadComplete** method receives a list that contains all
URLs of images. For the second phase, **LoadImages** uses a non-blocking operation
like **load**), then pushes the downloaded images into a **Queue**. **LoadImages** also
removes the URL of the image from a list (passed in Line 21), then the same
process is repeated until the URL list becomes empty. At the same time, the
display is invoked with a delay using a timer (lines 20 and 22). The **display**
method gets the downloaded image from **Queue**, then renders it on the circle.
The rendering images repeats until the the **Queue** becomes empty. As shown
in Figure 1, **LoadImages** and **DisplayPhoto** share a queue to pass/receive images.
Therefore, at a glance, it is not easy to understand the connection between
downloading and rendering of images because **execute** does not directly invoke
display using downloaded data.

   Moreover, **display** is implicitly invoked by a timer because the rendering of
images is faster than that of downloading. If this delay is not used, the repeat of

---

[1] We only show the classes required by the main behavior of FlickrSphere (*e.g.*, download and display images).

display unnecessarily consumes CPU resources. Although this delay is necessary, these pieces of code are difficult to understand at a glance. These pieces of code can be removed if the reference of **display** is passed to **LoadImages** then invoked inside a callback method (*i.e.*, **loadComplete**), making *callback spaghetti*.

```
1  class Searcher {
2    function search(key:String):void {
3      var loader:URLLoader = new URLLoader();
4      loader.addEventListener(Event.Complete, loadComplete);
5      loader.load(createURL(key));
6    }
7    function loadComplete(e:Event):void {
8      var list:Array = createList(e.data);
9      new DisplayPhoto().execute(list);
10   }
11 }
12
13 class DisplayPhoto {
14   private queue:Queue = new Queue();
15   private timer:Timer;
16   private loaded:Boolean = false;
17   function execute(list:Array) {
18     var loadImages = new LoadImages(queue);
19     timer = new Timer(2);
20     timer.addEventListener(TimerEvent.Timer, display);
21     imageLoader.load(list);
22     timer.start();
23   }
24   function display() {
25     if (queue.length == 0 && loaded) {
26       timer.stop();
27       timer.removeEventListener(TimerEvent.Timer, display);
28       return;
29     }
30     if (queue.length == 0) return;
31     var img:FlickrPhoto = queue.shift();
32     showImage(img);
33   }
34 }
```

Listing 1.3. A simplified implementations of FlickrSphere.

4   Applying Existing Proposals to FlickrSphere

This section presents different FlickrSphere implementations using existing proposals like async/await, Promise pattern, and SyncAS. While we present each implementation, we briefly explain each proposal.

4.1   The async/await Constructs

The async/await constructs is a proposal that supplies writing programs with non-blocking operations in a synchronous fashion for C# 5.0 [1]. A method invocation attached to **await** in order to keep the following executions as continuations that are restarted when the method execution is completed. The method usually contains an operation that takes a certain period of time (*e.g.*, **LoadImages.load**). Meanwhile, a method definition with **async** modifier lets the compiler know what the method contains a method invocation that uses non-blocking operations.

```
1  class Searcher {
2    void async search(String key) {
3      ListLoader loader = new ListLoader();
4      list = await loader.load(key);
5      new DisplayPhoto().execute(list);
6    }
7  }
8  class DisplayPhoto() {
9    LoadImages loader;
10   void execute(Array list) {
11     if (list) {
12       loader = new LoadImages();
13       display(list);
14     }
15   }
16   void async display(Array list) {
17     for (int i = 0; i < list.length; ++i) {
18       Image img = await loader.load(list.get(i));
19       if (img) showImage(img);
20       else i—;
21 } } }
```

Listing 1.4. Simplified main behavior of FlickrSphere with **async/await**.

Listing 1.4 shows the rewritten code using **async/await**. In Listing 1.4, we introduce **ListLoader** because **await** can be attached only to a method invocation that uses non-blocking operations and we assume **ListLoader** uses **URLLoader** inside the **load**. Using **async/await** enables writing pieces of code that contain asynchronous executions as synchronous fashion, making them easy to understand and intuitive.

4.2   Promise Pattern

One approach to deal with asynchronous issues adopted by JavaScript communities is the Promise pattern [3]: a proxy object that represents an unknown (or future) result that is yet to be computed. The common term used for promise is *thenable*, as a programmer uses a **then** method to attach callback methods to a promise when it is fulfilled.

```
1  class Searcher() {
2    function search(var key) {
3      loadList(key).then(new DisplayPhoto().display, error); // use a promise using then
4    }
5    function loadList(var key) { // create a promise object
6      var p = new Promise();
7      var loader = new URLLoader();
8      loader.addEventListener(Event.Complete, function(e) {
9        (e.data) ? p.resolve(e.data) : p.reject("error"); // choose which methods a promise invokes (success or fail)
10     });
11     loader.load(key);
12     return p;
13   }
14 }
15
16 class DisplayPhoto() {
17   var loader = new Loader();
18   var list;
19   function execute(list) {
20     this.list = list;
21     display();
22   }
23   function display() {
```

```
24      load(list.shift()).then(show, retry);
25    }
26    function load(url) {
27      var p = new Promise();
28      loader.addEventListener(Complete, function(e) {
29        (e.data) ? p.resolve(e.data) : p.reject(url);
30      });
31      loader.load(url);
32      return p;
33    }
34    function show(img) {
35      showImage(img);
36      if (list.length > 0) display();
37    }
38    function retry(url) {
39      list.unshirt(url);
40      display();
41    }
42  }
```

Listing 1.5. Simplified main behavior of FlickrSphere with Promise.

Figure 1.5 shows the rewritten code with the Promise pattern. Promise requires decomposing a set of operations into methods, then a method combines them creating a promise object and using **then**. Thereby, we create **loadList** (Line 5) to have a promise object and use it inside of the **Searcher.search** method. In the **DisplayPhoto** class, **load** is introduced, and **display** is rewritten to use the promise pattern. As a consequence, pieces of code that represent iterative **display** executions are non-intuitive because they use recursive invocations. Moreover, these recursions consist of a set of methods (*i.e.,* **display**, **show**, and **retry**), increasing its complexity. Note that, using loop statements such as **while** in **display** is impossible because it starts downloading and rendering all images at a time, leading a different behavior from the original FlickrSphere implementation.

4.3   SyncAS

SyncAS is a proof-of-concept library to provide *virtual block*, which enables a programmer to virtually block a method execution without blocking the execution of the program. A programmer specifies the points where a execution should be stopped and restarted using an aspect-oriented approach [6]. As a consequence, programmers can write programs as synchronous fashion even if they use non-blocking operations similar to **async/await**.

```
1  class Searcher {
2    function search(key:String):void {
3      var loader:ListLoader = new ListLoader();
4      var list:Array = loader.load(key); // a method invocation containing non-blocking operations
5      new DisplayPhoto().execute(list);  // virtually blocked by an aspect
6    }
7  }
8  class DisplayPhoto() {
9    private loader:LoadImages;
10   function execute(list:Array) {
11     if (list) {
12       loader = new LoadImages();
13       display();
14     }
15   }
16   function display(list:Array) {
17     var url:String = list.shift();
18     if (url) {
```

```
19    var img:FlickrPhoto = loader.load(url);      // a method invocation containing non-blocking operations
20    (img) ? showImage(img) : list.unshift(url);  // virtually blocked by an aspect
21    display(list);                                // self recursion
22  } } }
```

Listing 1.6. Simplified main behavior of FlickrSphere with SyncAS.

Listing 1.6 shows the rewritten code with SyncAS. Similar to async/await, SyncAS enables virtually blocking a method invocation that uses non-blocking operations. With SyncAS, we can write search in a synchronous manner without the need to add constructs like found in async/await. Instead, the ListLoader.load method is a method that contains a non-blocking operation, thereby, we need to deploy an *aspect* to virtually block the execution of Line 5 and restart them when loadComplete is finished as follows.

SyncAS.addAsyncOperation("ListLoader.load","ListLoader.loadComplete");

In addition, compared to loop constructs (*e.g.,* for) used in async/await, self recursive iterations are a bit non-intuitive, however, this recursion in SyncAS is easier than iterations over multiple methods used in Promise because this iteration can be naturally written using self recursion. To virtually block Line 19 in Listing 1.6, we need to deploy another aspect as follows.

SyncAS.addAsyncOperation("Loader.load","Loader.loadComplete");

4.4   Discussion

Applying each proposal can remove a timer, which connects LoadImages and Displayphoto as shown in Figure 1 and Listing 1.3, making pieces of code more intuitive. In addition, as shown in Table 1, we evaluate these proposals from three aspects: Modularity, Expressiveness, and Overload. *Modularity* refers to how we can concentrate one concern on one place. *Expressiveness* refers to how we can write programs naturally and intuitively. Finally, *Overload* refers to the difficulty that introduces each proposal.

Table 1. Comparison of proposals in terms of Modularity, Expressiveness, Overload

	async/await	Promise	SyncAS
Modularity	Middle	Middle	High
Expressiveness	High	Low	Middle
Overload	Thread level	Very low	Additional closure execution

The async/await proposal provides an appropriate solution that enables writing pieces of code as synchronous fashion. We can define a method that contains non-blocking operations at one place and use these constructs inside loop executions, thereby, its expressiveness is considered *high*. Since the base technique of this proposal is *Thread*, the overload is equivalent to Thread. Meanwhile,

async/await does not hide asynchronous executions completely because programmers, who just use a method containing non-blocking operations, need to have concerns about behaviors (*i.e.,* synchronous/asynchronous) in addition to its definition (*e.g.,* interface). This is because these programmers explicitly need to write **async/await** in order to control executions. As a consequence, the modularity of this proposal is not considered high (*i.e., Middle*).

The Promise pattern enables writing nested non-blocking operations at one place with a fluent interface using **then**. However, callback methods are necessary to follow the promise style, bringing modularity issues like callback spaghetti (*Low* expressiveness and *Middle* modularity). Moreover, iterative executions with non-blocking operations may bring complicated control flow (*e.g.,* recursive executions over methods) because understanding iterative and recursive executions is difficult at a glance. The overload is really *low* because this proposal is only a design pattern.

Although SyncAS has similar features to **async/await**, a SyncAS programmer, who uses a method containing non-blocking operations, does not need to have aware about behavior of methods. However, a programmer who provides asynchronous methods also needs to provide aspects that control asynchronous executions. This fact means that SyncAS is more modular than **async/await** and enables dividing programmers into two categories: non-asynchronous programmers who just use a method containing non-blocking operation, and asynchronous programmers who know and control asynchronous executions, enhancing modular development. Thereby, its modularity is the *highest* of these three proposals. Meanwhile, the current SyncAS forces programmers to write iterative asynchronous executions using self recursion instead of loop statements, leading to non-intuitive programs. Therefore the expressiveness is lower than async/await but higher than Promise (*Middle* expressiveness). As SyncAS requires the support of aspect-oriented programming that uses closure to weave an advice, its overhead depends on the *additional closure execution* time. In Listing 1.6 implementation, the additional execution time was 10[ms] while the downloading one image file took 500[ms], meaning that the effect of the additional time could be limited becuase asynchronous programming is usually applied to the execution that takes certain time.

Based on previous evaluations, a proposal that enables writing asynchronous executions as synchronous fashion enhances modular asynchronous Web programming. In this context, SyncAS is more modular than other two proposals, however, its lack of expressiveness and execution overhead is worse than other proposals. However its overhead is limited in practical usages because asynchronous programming is used when an execution takes a certain period of time. Introducing loop join point [5] and a sophisticated translator, which partially encapsulates programs related to asynchronous executions, are next challenges for current drawbacks.

5   Conclusion

Based on the progress in the Internet technologies, traditional applications are now provided as web applications. These modern web applications adopt asynchronous programming for various reasons: hiding latency of the network and improving the responsiveness in the user interface. Callback is a typical solution that enables asynchronous programming. However, this solution has its drawbacks such as *callback spaghetti* – the modern goto statement in asynchronous programming. In addition, introducing asynchronous programming into module based programming requires dividing a method into call-site and its continuations, making complex control flows. In order to solve these drawbacks, some proposals are available, however, issues related to modular programming, expressiveness, complexity are still present. In this paper, we evaluated and compared three proposals: async/await, Promise pattern, and SyncAS, applying them to a non-trivial open source application called FlickrSphere. From a modular programming view point, SyncAS is better than other two proposals because can encapsulate non-blocking operations in a module completely. From an expressiveness viewpoint, async/await is better due to supporting of loops (*e.g.,* for). Finally, the Promise pattern is only useful when developers need a lightweight solution. However, as we can appreciate in Table 1, none of these proposals fully support modular programming and expressiveness without adding a significant complexity.

Acknowledgments. This work was partially supported by JSPS KAKENHI Grant Number 26330089.

References

1. Bierman, G., Russo, C., Mainland, G., Meijer, E., Torgersen, M.: Pause 'n' play: formalizing asynchronous C♯. In: Noble, J. (ed.) ECOOP 2012. LNCS, vol. 7313, pp. 233–257. Springer, Heidelberg (2012). http://dx.doi.org/10.1007/978-3-642-31057-7_12
2. flickr: http://www.flickr.com/
3. Friedman, D., Wise, D.: The Impact of Applicative Programming on Multiprocessing. Technical report (Indiana University, Bloomington. Computer Science Dept.), Indiana University, Computer Science Department (1976). http://books.google.co.jp/books?id=ZIhtHQAACAAJ
4. Fukuda, H., Leger, P.: A library to modularly control asynchronous executions. In: Proceedings of the 29th Annual ACM Symposium on Applied Computing (SAC 2015). ACM Press, Salamnca, April 2015 (to appear)
5. Harbulot, B., Gurd, J.R.: A join point for loops in aspectj. In: Proceedings of the 5th International Conference on Aspect-oriented Software Development, AOSD 2006, pp. 63–74. ACM, New York (2006). http://doi.acm.org/10.1145/1119655.1119666
6. Kiczales, G., Irwin, J., Lamping, J., Loingtier, J., Lopes, C., Maeda, C., Mendhekar, A.: Aspect oriented programming. In: Muehlhaeuser, M. (general ed.) et al. Special Issues in Object-Oriented Programming (1996)

7. Mikkonen, T., Taivalsaari, A.: Web applications - spaghetti code for the 21st century. In: Proceedings of the 2008 Sixth International Conference on Software Engineering Research, Management and Applications, SERA 2008, pp. 319–328. IEEE Computer Society, Washington, DC (2008). http://dx.doi.org/10.1109/SERA.2008.16
8. Ogden, M.: Callback hell. http://callbackhell.com/
9. Parnas, D.L.: On the criteria to be used in decomposing systems into modules. Commun. ACM **15**(12), 1053–1058 (1972). http://doi.acm.org/10.1145/361598.361623
10. Yossy:beinteractive: Flickrsphere.   http://www.libspark.org/svn/as3/Thread/tags/v1.0/samples/flickrsphere/fla/FlickrSphere.html

Responding to Retrieval: A Proposal to Use Retrieval Information for Better Presentation of Website Content

C. Ravindranath Chowdary[(✉)], Anil Kumar Singh, and Anil Nelakanti

IIT (BHU), Varanasi, India
{rchowdary.cse,aksingh.cse,anil.nelakanti.cse}@iitbhu.ac.in

Abstract. Retrieval and content management are assumed to be mutually exclusive. In this paper we suggest that they need not be so. In the usual information retrieval scenario, some information about queries leading to a website (due to 'hits' or 'visits') is available to the server administrator of the concerned website. This information can be used to better present the content on the website. Further, we suggest that some more information can be shared by the retrieval system with the content provider. This will enable the content provider (any website) to have a more dynamic presentation of the content that is in tune with the query trends, without violating the privacy of the querying user. The result will be a better synchronization between retrieval systems and content providers, with the purpose of improving the user's web search experience. This will also give the content provider a say in this process, given that the content provider is the one who knows much more about the content than the retrieval system. It also means that the content presentation may change in response to a query. In the end, the user will be able to find the relevant content more easily and quickly. All this can be made subject to the condition that user's consent is available.

Keywords: Content management systems · Personalized content · Information sharing

1 Introduction

Information retrieval (IR) systems have become integral to daily activities of millions and will retain their prominence in years to come. One of the reasons for such importance of a good IR system is the amount of data that is available on the web and the pace at which it is increasing. The number of websites reportedly increased from one in 1991 to more than one billion in September 2014[1]. Simultaneously, there was an increasing number of users availing hosted services. This increase in web usage is more than an issue of load, that was met by computationally powerful servers. The bigger challenge was to organize and

---

[1] http://www.internetlivestats.com/total-number-of-websites/ on 27/10/2014.

© Springer International Publishing Switzerland 2015
F. Daniel and O. Diaz (Eds.): ICWE 2015 Workshops, LNCS 9396, pp. 103–114, 2015.
DOI: 10.1007/978-3-319-24800-4_9

make available the huge amount of information in a readily consumable manner. This required the third entity of retrieval systems. What essentially was a two-way transaction between the host and the client has become three-way with an IR system in the middle.

Clients are served by hosts, a relation facilitated by IR systems. However, current day IR systems are more than just organizers of web links. They model user choices and preferences to serve them better. We argue that the three-entity unit of the client, IR system and the host is greater than the sum of its parts. The relation between these three entities is ignored by the current web-service architecture. We present here a proposal which will exploit this relationship to better deliver some aspects for web service usage.

Web designers write content on the pages based on the information provided by the owner of the site. Content in a website is primarily organized based on the categorization of the information and arranged appropriately by the designer. In this entire process of the current design paradigm, the *query* has no role to play during the design or presentation phases of a website. But, when a search query is given to an IR system, it retrieves links of pages that are prepared without taking query into consideration *on the host side.* This is because retrieval and content management are considered mutually exclusive, that is, the content management system does not know about the retrieval system and the retrieval system does not know how the content provider may respond to the query. Due to this shortcoming, both the content provider and IR system are under-performing. In this paper, we try to address this issue by proposing an architecture that enables the server hosting the website to present content that is based on the query posed by the user.

In sections 2 and 3, we describe some background to our work and mention some related work. In section 4, we talk about query aware content presentation, which goes beyond conventional personalization. Section 5 is about some requirements that a proposed architecture should satisfy. The architecture itself is proposed in section 6. Section 7 is about Retrieval Response Protocol, with secion 8 giving a couple of examples to show how content presentation may change in response to the query. Section 9 is the conclusion.

2 Related Work

Authors Begen et. al. [3] have studied protocols for streaming video content both on the web and non-web. They show that web streaming, in contrast to the traditional broadcasting of video, requires various techniques to ensure good user experience. It is important to note that in the context of web streaming video content, there is often a role for interaction that includes video search and retrieval.

The relevance of links (and their content) to search queries and their ranking continues to be challenging, often requiring search with multiple attempts with variations of the query. Further, different results suggested by various search engines are also explored to find the required content. This complex behavior, particularly of experienced users, was studied by Bharat [4], who suggested keeping track of the queries with all their variations and the different results that we found relevant from the search engines that were used. This indicates how complex search and retrieval of information continues to be. It is in this context that we propose content provider's involvement in delivering the most relevant content from that site for a given query.

Research shows that content organization has a significant effect on how easy or complex the document is for the viewers to understand and consume. Document annotation has been suggested by numerous authors (see for example work of Ferri et. al. [9] or of Bottoni et. al. [5]) to capture, index and present content in an easy-to-understand manner on the web. We see our work as an extension to this wherein the content is customized to match user expectations using the query that led to it.

Personalized search is a an area that is gaining increasing interest in the research community to enhance user experience and deliver better results. The idea is to use cues beyond the search query, like the user profile, the search history, the context of the search, etc. to improve results. This goal was pursued in numerous works. Matthijs and Radlinski [13] model users using past browsing activity that includes the implicit feedback from clickthrough data. This information is then used to rerank the search results for a given retrieval task. Shen et. al. [15] also employ statistical language models along with the browsing activity exploited by Matthijs et. al [13]. Vallet et. al. [19] make clusters of related contexts by dynamically modeling the content from each retrieval. This is then used to discard irrelevant contexts for a given query and retrieve on contextually relevant content. Joshi et. al [12] give an overview of various methods used for this task in [9].

3 Feedback-Aware Retrieval

A classical or bare-boned retrieval system [6] only takes into account the query for retrieval. Some modern retrieval system go further and use the information available about the user for personalizing the results [1, 5, 9, 12, 13, 19]. Similarly, there has been work on Web search based on search context [4]. However, these systems do not take into account the user's activity once the user has selected and visited one of the web pages. In the proposed architecture, anonymized information about the user's activity will be made available to the retrieval system. It will, thus, be possible to design algorithms that take this activity into account. Some work in this direction was proposed by [15].

The details about this activity might include information such as the other links on the website that the user clicked on and the total time that the user spent on the website and on various pages. A retrieval system made aware of the feedback from the host server should, intuitively, perform better.

Some modern retrieval systems also provide additional links as part of the summary 'snippet' while presenting the results of retrieval to the user. Such snippets can be better prepared with the suggested feedback from the host server.

Additionally, and importantly, the host server can provide extra information about its content as part of the feedback. This extra information will be based on the query and the knowledge of the content that is available to the host server. This will allow the content provider to have a say in the presentation of the snippet for the concerned website. The retrieval system may or may not use this information, depending on the retrieval and snippet preparation algorithm.

4   Query-Aware Content Presentation

Current state of the art Web servers do not take the query into consideration while presenting the content to the user. Lot of work has been reported on improving the architecture of Web servers for various applications [3,11,20]. Many models are available to compare the architectures of the servers [10]. [22] discusses improving the performance of websites by using edge servers in Fog Computing Architecture. To the best of our knowledge there is no attempt to use the query by the host server to present the content to the user.

If a host server can present the content to the user based on the query, then it will be beneficial to both the user and the host organization. Suppose that user A gives query "popular movies in action genre + old" and that B gives "popular movies in action genre + latest". Let us assume that both the users get $Link1$ as their first link. We propose that the host server of $Link1$ should present different contents to each of them based on their query. In this case, the host server may present a list of old action movies (could be from other pages of the host server) to A and a list of new action movies to B, in both cases in addition to the content at the $Link1$.

For getting maximum benefit from this kind of architecture, the current Content Management Systems (CMS) like Drupal, Joomla, Django etc. may have to be redesigned to take user queries into account for presenting the final web page to be shown to the user. This will allow the host server (and the CMS) to play an active role in the process of content retrieval. Since the content provider knows much more about the content than the retrieval system, all that knowledge could be used to present dynamic query-aware content to the user.

While presenting the content to a user, web server can perhaps summarize the content that is available on its site and further increase the satisfaction of the user. There are many summarizers available to generate generic summaries [2,14]. Query-Specific summarization is discussed in [7,8,17,21]. A web server can go a step further and prepare a customized presentation to the user in line with [16,18]. The main point is that summarization can also be made specific to the user query on the retrieval system.

5   Privacy and Customization

Our proposal requires the retrieval system to share some information about the user and the query with the host server. It also requires the host server to provide feedback to the retrieval system based on the user's stay in the website after the user selected the website from the retrieval results. This extra sharing of information immediately raises the questions of privacy. If our proposal is implemented, its detailed version will need to include stringent requirements to address all the possible privacy concerns.

We list below some of these requirements:

- The first such requirement is that the user's identity, even if known to the retrieval system, will not be revealed to the host server. Whatever information is shared with the host server will have to be strictly anonymized so as not to reveal the user's identity.
- The second requirement is that only the relevant information will be shared. If we view this information as a list of attribute-value pairs, then only that subset of attribute-value pairs will be shared with the host server that the host server needs to know in order to better present its content.
- The third requirement is that an opt-out option will be available to both the user and the host server. The user will be made aware of the sharing of information and the user will decide whether this sharing is to be allowed or not. The information will be shared only if the user explicitly agrees to it. In the default case, there will be no sharing. Similarly, the host server will decide whether to provide feedback to the retrieval system or not and the default will be the latter.
- The fourth requirement is that both the user and the host server must be able to customize sharing of information. If they decide to share information, they will further be given the option to select the specific attributes that they are willing to share. For example, if the retrieval system knows about the user's location, age, gender and language, then the user may decide to share only location and language.
- The user will have to be informed that the activity on the visited website may be used for providing feedback to the retrieval system. And the user will then decide whether and what part of the activity on the website can be used to provide feedback to the retrieval system.

As this proposal is worked out in more detail in future work, more such requirements might be identified and will also have to be addressed.

Even after addressing these issues, one concern still remains regarding the proposed architecture. Even if the shared information is anonymized and the host server does not know the identity of the user, the retrieval system may still know the identity and be able to connect the activity of the user on the visited website with the user's identity. This raises the question whether the retrieval system will come to acquire more knowledge about the user than is warranted. This may be a problematic ethical issue and requires further investigation.

6   The Proposed Architecture

The outline of the architecture we propose is presented in Figure 1. The numbers used in this section refer to those in this figure. The scenario is that the user

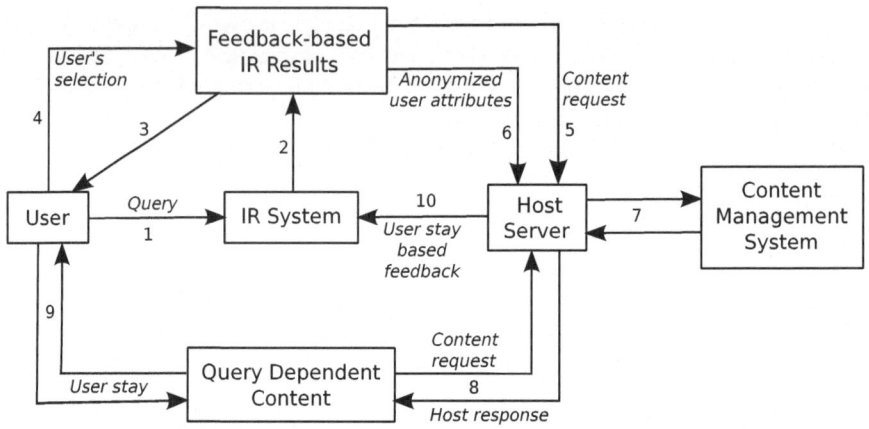

Fig. 1. The proposed architecture for more responsive IR and CMS systems

starts a retrieval system and gives a query (1). The retrieval system presents the search results to the user (2, 3). Out of them, the users selects one (4) and is taken to the destination website (5). When the user is taken to that website, the retrieval system also shares some information about the user and the query (subject to privacy requirements: see Section 5) with the server hosting the website (6). The host server uses this information to present the content (7) such that the user might have a better search experience (see Section 5). This presentation might, for example, make it easier for the user to find certain things. The host server will then provide feedback (10) to the retrieval system (again subject to privacy requirements) based on the user's stay in the website and the user's activity during the stay (8, 9). Since the information shared with the host server is anonymized, so will be the feedback given to the retrieval system. The retrieval system will now use this feedback to give better results in the future (see Section 3). The overall result will be better synchronization between the retrieval system and the host server for the purpose of presenting better results to the user. Anonymization, opt-out option and customization will be the central requirements, enforced through a protocol (see Section 7), to prevent any abuse that can result from sharing the information.

7   Retrieval Response Protocol

There are many different kinds of retrieval systems. Similarly, there are many different kinds of host servers and content management systems. If there is to be a flow of information between them as suggested in the preceding sections, then it will have to be precisely regulated so that it is possible to implement systems without any conflict. This will require a well-defined and well-designed protocol. We call this the Retrieval Response Protocol (RRP).

The Retrieval Response Protocol will regulate the flow of information between the retrieval system and the host server. The protocol will be used to initiate, maintain and close a *retrieval session*. As soon as the user selects one result from the results provided by the retrieval system in response to the user query, a retrieval session will be initiated. The ending of the session will perhaps have to be timeout based as there is no other way to know when the user has left the website.

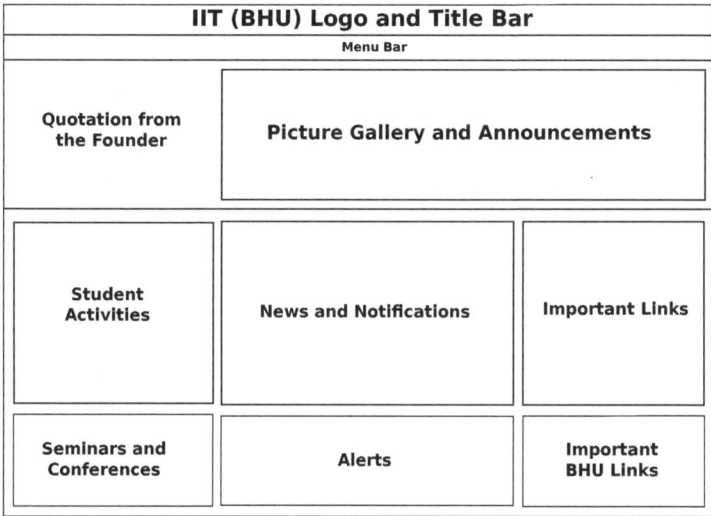

Fig. 2. Original page for the query 'IIT (BHU), Varanasi'

During the time the session is alive, the retrieval system will first share the information about the user and the query with the host server. After that, based on the user's activity, the host server will provide the feedback to the retrieval system. All the activity during this session will be subject to the privacy and customization requirements and the protocol design will have to take this into account.

The protocol will have to be designed to regulate this retrieval session. We leave the design of this protocol for future work.

8   Examples

In this section we show a couple of examples of how the content presentation by the host may change in response to the query in figures 2 to 6. We have taken two queries as examples:

– 'IIT (BHU) Faculty working on NLP information retrieval and text processing'
– 'Guest lectures in IIT (BHU)'

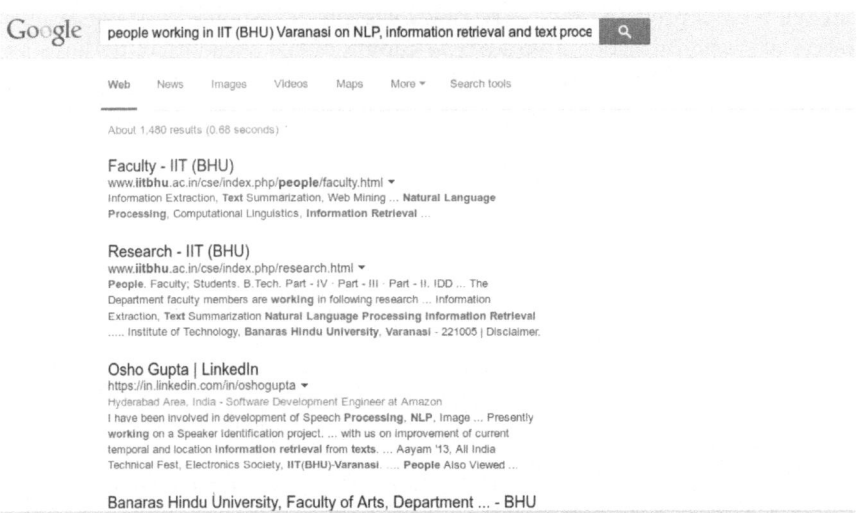

Fig. 3. First Google query results

IIT (BHU) Logo and Title Bar

Menu Bar

Quotation from the Founder	Picture Gallery and Announcements

Student Activities	**Faculty working on NLP, information retrieval and text processing** 1. Anil Kumar Singh ... 2. Ravindranath Chowdary C ... 3. Anil Nelakanti .... [Link to] News and Notifications	Important Links

Seminars and Conferences	Alerts	Important BHU Links

Fig. 4. Page presented in response to the first query

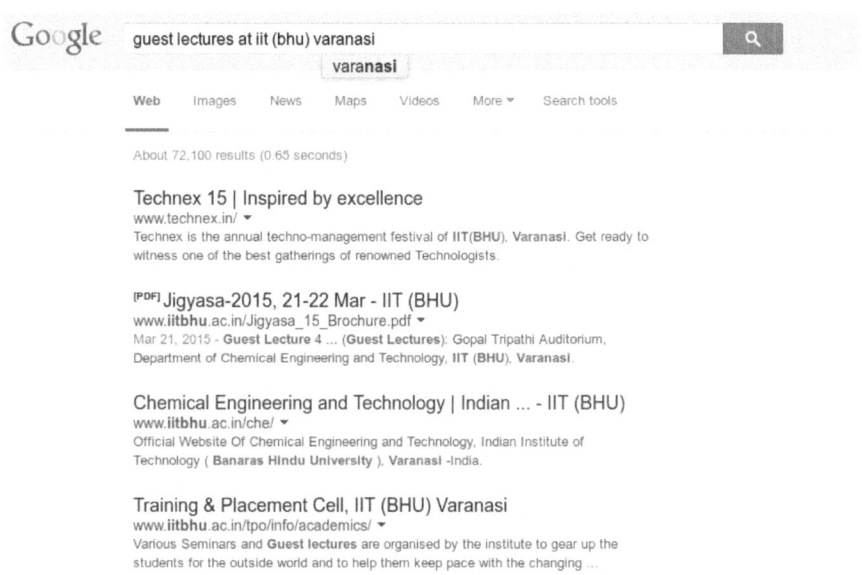

Fig. 5. Second Google query results

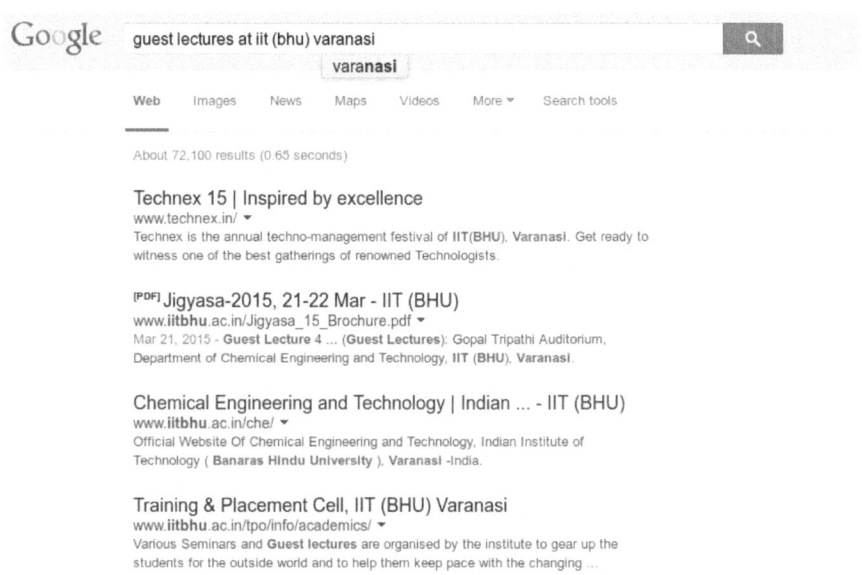

Fig. 6. Page presented in response to the second query

9 Conclusion

In the current information retrieval paradigm, the host does not use the query information for content presentation. The retrieval system does not know what happens after the user selects a retrieval result. And the host also does not have access to the information which is available to the retrieval system. We presented the outline of an architecture that addresses these issues. The aim is to provide a better search experience to the user through better presentation of the content based on the query and better retrieval results based on the feedback to the retrieval system from the host server. The retrieval system will share some information with the host server and the host server in turn will provide relevant feedback to the retrieval system based on the user's stay in the website. The host uses all the query related information for dynamic content presentation.

Most of this is not done as part of conventional personalization reported in the literature, where the host (with its content management system) does not really come into the retrieval picture except in just returning the pages requested. In conventional personalization, things are looked at only from the point of view of the retrieval system. One main idea here is to have varying presentation of the content based on the query (with the user's consent).

This revised paradigm for information retrieval also introduces the issues of privacy which will have to be addressed stringently. It also needs a new protocol for content retrieval response, which we briefly described. This protocol will regulate the flow of information between the retrieval system and the host server subject to the privacy and customization requirements. If some of what we propose is not feasible, say, due to privacy concerns, the rest could still be implemented.

We realize that the architecture proposed here is a bit sketchy. A direction for future research will be to flesh out the details of this architecture, including the retrieval response protocol.

References

1. Badros, G., Lawrence, S.: Methods and systems for personalised network searching (2005). US Patent Application 20050131866
2. Balahur, A., Kabadjov, M.A., Steinberger, J., Steinberger, R., Montoyo, A.: Challenges and solutions in the opinion summarization of user-generated content. J. Intell. Inf. Syst. **39**(2), 375–398 (2012)
3. Begen, A., Akgul, T., Baugher, M.: Watching video over the web: Part 1: Streaming protocols. IEEE Internet Computing **15**(2), 54–63 (2011). doi:10.1109/MIC.2010. 155
4. Bharat, K.: Searchpad: explicit capture of search context to support web search. In: Proceedings of the 9th International World Wide Web Conference, pp. 493–501 (2000)
5. Bottoni, P., Ferri, F., Grifoni, P., Marcante, A., Mussio, P., Padula, M., Reggiori, A.: E-document management in situated interactivity: the wil approach. Universal Access in the Information Society **8**, 137–153

6. Brin, S., Page, L.: The anatomy of a large-scale hypertextual web search engine. In: Proceedings of the Seventh International Conference on World Wide Web 7. WWW7, pp. 107–117. Elsevier Science Publishers B. V., Amsterdam (1998). http://dl.acm.org/citation.cfm?id=297805.297827

7. Chowdary, C.R., Kumar, P.S.: ESUM: an efficient system for query-specific multi-document summarization. In: Boughanem, M., Berrut, C., Mothe, J., Soule-Dupuy, C. (eds.) ECIR 2009. LNCS, vol. 5478, pp. 724–728. Springer, Heidelberg (2009)

8. Chowdary, C.R., Sravanthi, M., Kumar, P.S.: A system for query specific coherent text multi-document summarization. International Journal on Artificial Intelligence Tools **19**(5), 597–626 (2010)

9. Ferri, F., Grifoni, P., Padula, M.: Using shape to index and query web document contents. Journal of Visual Languages and Computing **13**, 355–373

10. Harji, A.S., Buhr, P.A., Brecht, T.: Comparing high-performance multi-core webserver architectures. In: Proceedings of the 5th Annual International Systems and Storage Conference, SYSTOR 2012, pp. 1:1–1:12. ACM, New York (2012). http://doi.acm.org/10.1145/2367589.2367591

11. Hashemian, R., Krishnamurthy, D., Arlitt, M., Carlsson, N.: Improving the scalability of a multi-core web server. In: Proceedings of the 4th ACM/SPEC International Conference on Performance Engineering. ICPE 2013, pp. 161–172. ACM, New York (2013). http://doi.acm.org/10.1145/2479871.2479894

12. Joshi, C., Jaiswal, T., Gaur, H.: An overview study of personalized web search. International Journal of Scientific and Research Publications **3** (2013)

13. Matthijs, N., Radlinski, F.: Personalizing web search using long term browsing history. In: Proceedings of the Fourth ACM International Conference on Web Search and Data Mining, pp. 25–34. ACM, February 2011

14. Radev, D.R., Jing, H., Budzikowska, M.: Centroid-based summarization of multiple documents: sentence extraction, utility-based evaluation, and user studies. In: NAACL-ANLP 2000 Workshop on Automatic Summarization, pp. 21–30. Association for Computational Linguistics, Seattle (2000)

15. Shen, X., Tan, B., Zhai, C.: Context-sensitive information retrieval using implicit feedback. In: Proceedings of the 28th Annual International ACM SIGIR Conference on Research and Development in Information Retrieval. SIGIR 2005, pp. 43–50. ACM, New York (2005). http://doi.acm.org/10.1145/1076034.1076045

16. Shibata, T., Kurohashi, S.: Automatic slide generation based on discourse structure analysis. In: Dale, R., Wong, K.-F., Su, J., Kwong, O.Y. (eds.) IJCNLP 2005. LNCS (LNAI), vol. 3651, pp. 754–766. Springer, Heidelberg (2005)

17. Sravanthi, M., Chowdary, C.R., Kumar, P.S.: QueSTS: a query specific text summarization system. In: Proceedings of the 21st International FLAIRS Conference, pp. 219–224. AAAI Press, Florida, May 2008

18. Sravanthi, M., Chowdary, C.R., Kumar, P.S.: Slidesgen: automatic generation of presentation slides for a technical paper using summarization. In: Proceedings of the Twenty-Second International Florida Artificial Intelligence Research Society Conference, Sanibel Island, Florida, USA, May 19–21, pp. 284–289 (2009). http://aaai.org/ocs/index.php/FLAIRS/2009/paper/view/22

19. Vallet, D., Fernandez, M., Castells, P., Mylonas, P., Avrithis, Y.: Personalized information retrieval in context. In: Proceedings of 3rd Int. Workshop Modeling Retrieval Context 21st Nat. Conf. Artif. Intell. (2007)

20. Veal, B., Foong, A.: Performance scalability of a multi-core web server. In: Proceedings of the 3rd ACM/IEEE Symposium on Architecture for Networking and Communications Systems. ANCS 2007, pp. 57–66. ACM, New York (2007). http://doi.acm.org/10.1145/1323548.1323562
21. Yin, W., Pei, Y., Zhang, F., Huang, L.: Query-focused multi-document summarization based on query-sensitive feature space. In: Proceedings of the 21st ACM International Conference on Information and Knowledge Management. CIKM 2012, pp. 1652–1656. ACM, New York (2012). http://doi.acm.org/10.1145/2396761.2398491
22. Zhu, J., Chan, D., Prabhu, M., Natarajan, P., Hu, H., Bonomi, F.: Improving web sites performance using edge servers in fog computing architecture. In: IEEE 7th International Symposium on Service Oriented System Engineering (SOSE), pp. 320–323, March 2013

Perspectives and Methods in the Development of Technological Tools for Supporting Future Studies in Science and Technology

Davide Di Pasquale$^{(\boxtimes)}$ and Marco Padula

ITC-CNR, Construction Technologies Institute, National Research Council, Milan, Italy
{davide.dipasquale,marco.padula}@itc.cnr.it

Abstract. The term "future studies" refers to studies based on the prediction and analysis of future horizons, able to examine the long-term impact of policies and technologies, and to anticipate emerging social challenges. Those studies experience today an emerging interest, also due to the large sets of data made available by the social media and big data phenomena. This paper presents a review of widely adopted approaches in these study activities, with three levels of detail scale: starting from a wide scale mapping of related disciplines, the second level focuses on the traditionally adopted methodologies and, on the level of greater detail, the paper describes methodological and software analysis tools that cope also with the semantic web and social media aspects.

The paper finally proposes the architecture of an extensible and modular support platform for these study activities, able to offer and integrate a number of tools and functionality oriented to the harmonization of aspects related to semantics, document warehousing and social media aspects.

Keywords: Future Studies · Foresight 2.0 · Collaborative tools · ICT platforms

1   Introduction

The term "future studies" refers to studies based on the prediction and analysis of future horizons, able to examine the long-term impact of policies and technologies, and to anticipate emerging social challenges. Those studies represent a discipline with a deep-rooted practice in specialist areas since the 50s that today experience an emerging interest, also due to the large sets of data nowadays made available by the exponential growth of social media and big data phenomena. The field of forecasting includes the study and application of judgment as well as of quantitative or statistical methods and by now represents an important phase of management decision making in all sectors.

One of the fundamental assumptions of Future Studies is that there is a plurality of alternative futures of varying probability and the object of study is therefore to identify and describe alternative futures by collecting and processing qualitative and quantitative data about the possibility, probability and desirability of change. The scholars

F. Daniel and O. Diaz (Eds.): ICWE 2015 Workshops, LNCS 9396, pp. 115–127, 2015.
DOI: 10.1007/978-3-319-24800-4_10

of the discipline have recently begun to examine social systems and to build scenarios, discussing worldviews behind them through methods of stratified causal analysis.

This paper aims to contribute to the empowerment of methodological approaches to future studies through the use of the Information and Communication Technologies and is structured into two main parts. The first part analyses the approaches adopted in these study activities, presenting a review of the background context with three levels of detail scale: in the first subsection it presents a wide scale mapping of related disciplines, in the second it introduces the most used methodologies and in the third subsection, on the level of greater detail, the paper describes techniques and software analysis tools that cope also with the semantic web and social media aspects.

In the second part the paper proposes the architecture of an extensible and modular support platform for these study activities, able to offer and integrate a number of tools and functionality oriented to the harmonization of aspects related to semantics, document warehousing and social media aspects.

2   Background of Futures Research

The Academic interest in Futures Research in Europe has been influenced by the concepts of participation and collaboration, through the idea of a narrative and non-deterministic approach to socio-economic forecasting, authored by "previsionary forums" where "experts from very different disciplines contribute their individual foresight, resulting in harmonized depictions about possible futures" (de Jouvenel, 1967). Their goal is to study social developments, behaviors and processes to advance and enable collaborative action upon the future. Until today, those ideas have a profound influence on common methods of Futures Research. (Bell, W., 2002).

In the following paragraphs a review of disciplines related to the study and forecast of possible future scenarios is presented along with the techniques and tools most widely adopted to approach this study domain.

2.1   The Context of Related Disciplines

In the context of future studies a number of related disciplines approach different aspects of the study domain, focusing on different time span extension or different ways of organizing data collection or differentiating on the way they try to predict, prospect or pro-actively regulate the vision they are exploring. Hereafter it follows a brief description of some of these fields of study.

Strategic Planning a normative method by definition used as a top-down ap-proach, evokes the creation of a plan as a method of getting from one set of circum-stances to another. It commonly consists of a formulation of a whole system picture, then refined with reiterated passages. It tends to focus on shorter-term, more predicta-ble topics compared to foresight studies and take the objectives and aims of the activi-ties as ends for which we are seeking to define the most effective means (van der Heijden, 2005).

Scenario Planning can be defined as an internally consistent view of what the fu-ture might turn out to be (Porter, 1985) and scenario planning is considered that part of strategic planning which copes with technologic aspect related to future uncer-tain-ties (Ringland, 1998). The study of uncertainty in its social context is thought to pro-duce understanding and more robust long-term policies (Chermack, 2005). It consist of a qualitative approach to the generation of structured data in order to create multi-ple potential future scenarios that reflect cyclical uncertainties. Making use of creative and narrative approaches it helps to synthesize and learn about the context within which specific policies will unfold. Scenarios studies can be classified in three main categories with 2 sub types each (Börjeson et al., 2006): *predictive* (*Forecasts* and *What-if* sub-types), *explorative* (*External* and *Strategic* sub-types), *normative* (*Pre-serving* and *Transforming* sub-types). The scenario planning process usually includes the following steps: 1. Identify the issue, 2. Identify the key factors, 3. Re-search driv-ing forces, 4. Rank key factors and driving forces, 5. Develop scenario logics, 6. De-velop scenario details, 7. Consider implications, 8. Identify indicators.

Horizon Scanning copes with exploring the external environment to better under-stand the nature and pace of change in that environment, and identify potential oppor-tunities, challenges, and likely future developments, providing a basis for analysis of future program and decision-making (Conway, M., 2000, Palomino et al., 2013). It can be split into Evidence-based Horizon Scanning (Deductive approach, usually static, periodic, and issue focused), and Intelligence-Based Horizon Scanning (Induc-tive approach, usually dynamic and continuous).

Foresight and Beyond. The term foresight is used to distinguish from forecasting and emphasize its explorative nature. The scope of the term foresight extends to both normative and explorative approaches to Futures Research. Strategic foresight copes with reducing the domain of the unknown and considering uncertainty in the decision-making process. Methods have to be chosen and tailored to fit the specific task com-bining quantitative and qualitative approaches (Heger & Rohrbeck, 2012) and two main forms of foresight linked to policy can be described. There is foresight for/in policy, relating to its advisory and strategic function, where foresight serves as a tool to inform and develop policy in any area or to "join up" policy across domains. An-other approach considers foresight relating to its instrumental role, where it serves as an instrument to implement budgetary, structural or cultural changes in the domain of research and/or innovation policy (Cassingena et al., 2008). Foresight processes have a strong link with innovation policy as they provide the means to guide, develop and shape innovation systems or sectors. These processes also serve as an instrument to improve the effectiveness of innovative public procurement, industry-university links, cluster policy, and the development of technology platforms. So their benefits are both strategic and instrumental.

The large-scale adoption of internet access that happened over the last 15– 20 years let emerge the potential to include several hundreds of participants remotely in a real-time foresight experiment, opening a novel perspective on how foresight processes change qualitatively and quantitatively when they are built on the backbone of a large-scale IT infrastructure (Schatzmann et al. 2013; Pang e al., 2010). According to Schatzmann this novel paradigm comprehends using online frameworks and a mas-sively collaborative approach, leading to the concept of *open foresight*: it can be de-scribed as context-based foresight as it is typically based on sharing and collaboration upon all steps in a foresight process, from scenarios to future strategies. Particular emphasis is given to the integration of sociocultural and technological dynamics that arise with a ubiquitously connected society. In modern foresight exercises, online communities can be considered as a participated *think-tank* in which future concepts, ideas or scenarios can be tested and refined.

2.2   Widely Adopted Methodologies

Methodologies commonly adopted to deal with the different stages of the study activity can be roughly divided into two categories, based on the type of approach that characterizes them: qualitative or quantitative. Among those based on a qualitative approach we can include the following techniques.

Intuitive logic approach is a workshop-based approach often used for the development of the scenarios, involving stakeholders and representatives of the using focus groups and focal questions as building blocks of the scenarios (Carlsen et al. 2013). The participated approach improves the quality of the scenarios and also contributes to their acceptance among the users (van der Heijden 2005).

The core of **Delphi** technique is to collect and harmonize the opinions of a panel of experts on the specific topic of interest. It considers the judgement of a selected number of informed people as the optimum starting point to forecasts (Börjeson et al., 2006). In the original Delphi method, questions are sent to a panel of experts in various rounds, followed by synthesis and debate; the result is a consensus forecast or judgement.

The **Alternative futures analysis** approach is usually adopted in urbanistic and naturalistic studies; it makes use of GIS infrastructure and provides visualization and community assessment of alternatives to balance political and natural science constraints (Steinitz, 2003). It's particularly adopted in environmental assessment studies helping communities make decisions about land and water use (Environmental Protection Agency, 2002).

Backcasting is defined as an actual backward-looking analysis, rather than an overall process. This technique usually involves a group discussion process that first creates a desirable future vision or a normative scenario, then tries to determine how to realize this desirable future defining and planning activities and strategies leading towards its realization. The methods adopted are (as described in Kok et al., 2011): 1) define the end point vision; 2) indicate obstacles to be overcome and opportunities to be taken; 3) define milestones and interim objectives; 4) identify policy actions and specify involved actors; 5) identify main drivers and weaknesses.

The **SWOT Analysis** refers to the acronym and technique used to evaluate (s)trengths, (w)eaknesses, (o)pportunities and (t)hreats involved in a process. Identification of such aspects is obviously important in each forecasting activity as they can help in developing strategies and planning steps to achieve the objectives.

Among the methodologies based on a more **quantitative approach** we can include the trend impact and the cross impact analysis.

The **trend impact analysis** (Gordon et al., 2003) is a method for adjusting the baseline trend given the occurrence of a potential future event. It takes in consideration a perturbation event that possibly affects a given trend and tries to evaluate the resulting modified scenarios. Three different scenarios are evaluated, corresponding to three notable events: the first noticeable effect of perturbation, the maximum per-turbation effect and the final result of the event. A new trend line is then calcu-lated as an alternative scenario and compared to the original situation.

The **cross-impact analysis** technique introduces in the forecasting process a ma-trix of estimates of event interdependence (Smith, 1987). It enables a more quantita-tive approach to scenario development and often involves the analysis of real world sce-narios to determine a baseline against which produce high, medium, and low va-ria-tions through expert and stakeholder surveys or workshops. Probability distribu-tions are estimated for groups of trends and variables, providing a more integrated numeri-cal estimation of possible futures (Bradfield et al., 2005).

2.3   Methodological Techniques and Software Analysis Tools

Developments in ICT have produced a range of alternatives commonly referred to as *Web 2.0* approaches that include: individual production of user-generated content, including amateur contributions, *Folksonomic* (Vander Wal, 2005), tagging i.e. user-generated annotation of data, shared with the community, data aggregation and social filtering, participation and openness in terms of data, API's and intellectual property.

Typically foresight studies in support of long-term visions have been conducted over the last decades by cities, regions, nations and world regions, often making use of internet tools for questionnaires, Delphi studies or online assessments for stake-holder engagement. In contrast with these more traditional ways of stakeholders' involvement in foresight studies, social web 2.0 platforms introduce real time feed-back into the process, allowing for more interaction and collaboration. Stakeholders can see each other's responses in real time and react on them.

Hereafter a brief outline of frequently adopted operative techniques and software tools for supporting future studies activity are described.

Crowdsourcing is a term that describes the outsourcing of a function or activity to an undefined (and generally large) network of people in the form of an open call (Howe, 2006). The gathering of massive number of participants into a bottom up ex-ercise can be started through a call for participation open to the citizenship or to a subset with particular characteristics.

OLAP is the main paradigm for accessing multidimensional data in data ware-houses, providing operations to transform one multidimensional query into another. An OLAP service meets the following requirements and characteristics: the base level of data is summary data; availability of historical, current, and projected data; aggre-gation of data and the ability to navigate interactively to various level of aggregation (drill down); derived data which is computed from input data; multidimensional views of the data; ad hoc fast interactive analysis (response in seconds); medium to large data sets (1 to 500 Gigabytes). This software tool was recently applied to docu-ment and text mining applications (Drzadzewski et al., 2012, Zhang et al., 2009) and is needed in case of large dataset production when the hidden causal links among enti-ties must be accurately examined with data mining approach (Fang, 2010).

Ranking algorithms are used to obtain a quantitative measure of the affinity of a generic item (product, term search, document) in an unattended procedure. Collabora-tive filtering has been used to predict the rating a target user assigns to an item (Koren et al. 2011, Shuaiqiang et al., 2014), for example using the ratings on the item as-signed by other users who share similar rating preferences or using its correlation with other similar items through matrix factorization techniques (Schafer et al., 2001). These algorithms are the base of recommendation system included in many web services and facilities.

Recommendation systems have been widely used at e-commerce websites to identify from a huge range of products the most appropriate ones for each user. Various recommendation methods have been proposed to predict the wanted products based on users' preferences as well as preferences of similar users or finding the most appropriate people to follow, tweets to read, as well as hashtags to use in their tweets or documents to read (Gipp et al. 2009). These system are based on one or a combination of many ranking algorithms. There are two main types of recommendation systems – personalized and non-personalized, depending on the fact that the adopted ranking algorithm takes into account the specificity of each user. Recommenders can be used to help the development of crowd-sourced discussion activities suggesting pertinent contents based on keyword similarity or user affinity (Pham, 2011).

In the context of foresight 2.0, **Social Rating Systems** are used for various purposes, mainly to collect and rate base data like trends for deriving predictions or to collect and rate predictions and test their likelihood or desirability (Su et al., 2009). After collecting any assumption or prediction, they can be also rated on scales like "rele-vance" or "impact against a set of specific users or experts.

Planning Support Systems are planning tools that exploit ICT to aid planning efforts in a way which go "beyond geographical information systems" (Brail and Klosterman, 2001). These platforms link territorial informative systems to policy making activities, enabling decision-makers with specific toolbox and. These include Public Participation GIS efforts (PPGIS), that enhance the availability of GIS to the public like Web-based GIS facilities, soliciting opinion and incorporating local knowledge, and participatory agent-based modelling exercises, that use bottom-up stochastic agents and social simulation and have been successfully used as education-al and empowerment tools, as well as for policy impact assessments.

Foresight Support Systems' main goal is to support the complex activity of futures study with tools that enable collaboration, ensure transparency and consistency of foresight results and support the efficient handling of very large volumes of data (Ra-ford, 2014). This can be achieved without limiting the flexibility of the foresight process and, according to Banuls and Salmeron (Durst et al., 2014), supporting the rules of order in foresight processes, the combination and integration of different foresight methods, the reuse of results of foresight activities, and the collaborative decision-making. This kind of platform has to offer a wide range of foresight methods integrated into a seamless foresight process covering individual and collaborative methods. These systems often provide project management functionalities and a database to store the results of all foresight activities; decision-makers, experts and partners from other organizations are enabled to collaborate in shared projects with their specificity in terms of expertise but also taking into account information access rules and permissions. These ICT platforms have a variety of specific functions and enabling facilities and are of extreme interest as collectors of novel approaches to ICT powered future studies (Rohrbeck et al., 2013). In the next paragraph a detailed description of a possible architecture of an informatics platform ecosystem devoted to future studies activity is proposed.

3   An Extensible and Modular Platform Architecture for Supporting Futures Study Activities

According to (Haegeman et al. 2012) ensuring engagement and participation of stakeholders depends partially on the platform design. Aspects to consider inclulde: *registration* - most platforms require registration before participants can contribute, this allows identifying participants; *differentiation* between roles that can be played in the platform and the kind of information that is accessible; *customization* of information displayed to participants; *simple design*, factor pushing the intensity of usage.

Hence any project that aims to build an ICT tool able to perform forward-looking analysis to outline scientific futures and "reliable" scenarios of scientific and technological development should consider these aspects in the design phase, starting from the target approach chosen to drive the initiative.

One possible approach aims at supporting a community of experts which are addressed to a limited set of focus groups each identified by a selected topic. Participants in the community have to be enabled to share knowledge, to identify gaps in knowledge, to point out obstacles, to identify needs for more and better education as well as for more funding, and to outline market potential and social acceptability for activities and products. A web-platform that represents the operational infrastructure of the collective network of the process must enable and enhance knowledge sharing between different competences, integrating a number of different modules each focused on a specific aspect of the workflow:

- a Library module: to organize and catalogue the data sources used for the produc-tion of documents and to store resources for studies;
- a Discussion module: to allow the exchange of ideas and brainstorming activities, collect members discussions thematically organized and moderated so that to potentially be used as a basis for the production of new documentation;
- a Knowledge management module: to collaboratively author documents, basing on material from the discussion module and supported by library module's items.

Another possible approach is aimed at relating wide audience of users / citizens: an ICT infrastructure to support this activity must therefore offer social and collaboration tools that allow citizens to express, share, develop and then evaluate ideas about future scenarios that that would like to see implemented. This workflow is supported by a collaborative platform comprising the following elements:

- a front-end section that provides social and collaborative tools such as wikis, voting systems and tools for the management of documents such as scientific articles, foresight reports or Web resources to which it is possible to add semantic annota-tions in order to improve utilization and visibility;
- a set of tools for knowledge warehousing, available to both citizens and policy makers; combining statistical analysis with tools for data mining and simulation environments, they facilitate the management of acquired and produced information, as well as the identification and the understanding of relationships and connections between them;
- a set of software agents specialized in collecting information from mainstream social networks as well as specialist web-available contents: these acquired infor-

mation can be also used to assess the emotional approach of individuals to fore-sight issues.

Considering the context of European Research, an example of such an ICT plat-form can be found in the Futurium platform used by Digital Futures, a foresight pro-ject launched by the European Commission's Directorate General for Communica-tions Networks, Content and Technology (Accordino, 2013). Its architecture consists of the following components: front-end participatory tools, knowledge-harvesting tools for both policymakers and stakeholders, data-crawling tools to extract knowledge from popular social networks and embed it into the platform and data-gathering tools to fetch real world data. It offer several features to support participa-tory foresight: co-creation of futures and policies (wiki), voting system about futures and policies, annotated digital library for referencing document production.

While considering the divergence in the foresight vision and in the content of the two approaches, however they converge on the mission and on the issues they face; they should also share the basic structure of resources in order to allow qualified and di-versified access to users, the organization of content in a common space that allows the deepening on the topics of interest, a list of keywords for structured indexing and data integration.

Table 1. comparison of ICT platforms supporting futures studies activities

	first approach	**second approach**
actors	community of experts	wide audience of users / citizens
organization	focus groups each identified by a selected topic	social community
needs	share knowledge, identify gaps in knowledge, identify obstacles, identify needs, outline market potential	express, share, develop, evaluate ideas
main func-tionality	knowledge sharing	social comunication, collabora-tion
tools	library, discussion forum, knowledge management	wiki, voting system, management of documents, knowledge ware-house, agents for collecting info

Table 1 collects and compares the main features of two platforms supporting fu-tures studies activities; futures studies can be very complex due to the contexts and prob-lems they face, the high quantity of variables to be considered and their possible un-certainty; platforms and ecosystems adopted must satisfy the needs of the actors in-volved, fit the activities that are chosen to be supported and the results which have to be reached. Very often ICT platforms are specifically developed as, for example in (Rohrbeck et al., 2013; Durst et al., 2014); we believe that there is no way to choose an optimal environment due to the intrinsic variability of the futures studies; it would be more effective to develop a repository of specific components which can be select-ed and integrated into the kernel of a general purpose platform to configure it.

Platforms Roles and Ecosystems

Infrastructures based on both these approaches can and should co-exist, given their complementary role, becoming a holistic tool for knowledge sharing and knowledge management for communities of researchers or, in general, online users (Grifoni et al., 2014). As examples of such ecosystems it can be cited the platform implemented for the project SHAPES. It implements a specific framework (PLAKSS) that inter-connects data, services, applications, and knowledge in a mash-up perspective, and integrates web services and social networks into a common environment that aims to enable the building of personalized and structured knowledge bases selecting data and information from available resources, categorizing and integrating them in new knowledge (Ferri et al., 2014).

Considering the examples described, particularly the European Futurium platform, a possible specificity can be identified declining nationwide the specialist platforms (described above as top down approach) and converging appropriately filtered outputs of these national exercises on the supranational European bottom up approach platform, where the likeness and appeal of the produced scenarios can be tested against citizens. Actions aiming to establish a systematic exchange of information between the two types of platforms should consider the following points:

- The collaborative model / ecosystem to be realized must provide a dual flow of contributions between the supranational platform and the different national platforms, each specialized both as regards the type and the final users of the produced knowledge. The desirable vision consists of an ecosystem of platforms supporting national foresight initiatives of the 28 member countries. This ecosystem is charac-terized by the exchange network between the various entities that make it up, and between them and the external environment, and we can identify different actors and outline the relationships between them: users who use different platforms, software artefacts that underlie the existence of the ecosystem and the flow of in-formation that is both at abstract level (knowledge) and practical (data flow);
- The interoperability between platforms must be provided with a content-based model, providing strategies and tools for standardization, processing, cataloguing and sharing content at different levels of structuring and complexity, from raw data to complex scenarios;
- The access to the contents has to be provided by means of tools that allow high granularity in access policies, so as to precisely determine which types of users can access what types of content and the different eligible manipulation operations on such contents (privacy or copyright issues);
- National and supranational dimension: on the one hand the flow of specialist con-tributions from the various national scientific communities feeds discussion and crowdsourced creativity provided by the European platform, on the other the ideas and visions promoted and catalysed from European citizenship, fuels discussion and research topics on a national scale.

Such a supranational platform can then carry out, on the one hand, the role of intro-duction and dissemination in the population of information on new scenarios and new technologies and, on the other, can provide to decision makers (at national and Euro-pean level) assessment tools about the address policies considered as priorities by citizens and / or specialists.

Interoperability and Integrating Tools

The interoperability between the supranational and the national platform is needed to ensure the automation processes of content mining and retrieval of research results by users with different features and privileges. It should also ease the interchange of data sources and possibly permitting the preparation of common repository accessible by the entire ecosystem.

Prerequisite to semantic integration of contents present on different platforms is that there is a common or at least analogous semantic structure that can be effectively exploited for cataloguing the contents. Classification using authors' keywords is certainly accurate but nonetheless also poses the problem of using shared or "interoperable" taxonomies. This problem can be addressed according to three approaches:

- mapping: a translation is made of each taxonomic node establishing a relationship between elements of each set;
- merging: a common and shared taxonomy is created merging the existing ones;
- alignment: relations between subsets of nodes of the original taxonomies are creat-ed in order to set up a translation matrix among them.

The latter approach has the advantage that does not require changes to the original taxonomic structures and to allow the coexistence of different alignment matrices (one for each pair of taxonomies to interface).

Operationally, this step can be carried out with the use of techniques of Ontology mapping, working both on the merely conceptual plane (the formal representation of concepts), and on the "social" side, that is, taking into consideration the relationships between individuals seen as agents of the cataloguing system.

The communicational interoperability between platforms can be developed through a layer of exchange that must be transparent and present a specific interface for each system used by national platforms, and must provide a consistent interface to the out-side. The technical implementation of this level can be obtained through the use of application programming interfaces (APIs) that can expose Web services in the common paradigm of web services. Thus, once defined eligible operations using XML descriptors (WSDL), you can automate the processes of research, adaptation and presentation of content from one platform to another.

The Semantic Management of Resources

The semantic structures used to manage resources in the two types of platforms can be enhanced by introducing or improving a taxonomy of keywords dynamically fed according to a top down specialist approach, characterized by marking the contents "ex ante" (in the authorial phase), with expert driven taxonomies and / or thesauri and / or specific vocabularies, or according to a bottom up participated approach, charac-terized by marking "ex post" content by users / readers with free terms (folksonomy) or by applications (semantic mining). Among these we distinguish between super-vised mode (taxonomies: vocabularies with hierarchical structure; thesauri: shared vocabularies and controlled semantic structure, glossaries and unstructured sets of keywords defined in specific initiatives; folksonomies: teamed annotations by users) or automated (semantic mining: using algorithms and ontologies to automatically classify content).

Proposed Tools

The use of such annotation and semantic cataloguing systems allows for example to introduce in the platforms specialized tools for the retrieval of related content from the supranational platform. These tools can comprehend:

- In-context searches for related content in other platforms: exploiting the key-words annotation used in the process of contribution authoring and automating content discovery on other platforms;
- Recommended links: a tool that allows to manually create and maintain a list of links to content considered useful to a specific topic of discussion. Such information remains associated with individual topics, encouraging indexing and cataloguing;
- Structured presentation of content: a tool that presents in a semantically organized manner the content retrieved automatically from ecosystem's platforms;
- Suggested related terms: a tool that analyses the annotated taxonomy of keywords used in a discussion/contribution and shows the related keywords that are associat-ed with similar contents on other platforms. These keywords can suggest combina-tions of semantic-related terms.

4   Conclusions and Future Works

ICT had shown and is showing a rapid evolution driving the change of habits in all spheres of social life within our society, not least in research activity. Many tools are yet available for different aspects of futures and foresight studies, supporting their different phases, from the remote interaction of experts' panels to the implementation of questionnaires and elaboration of their results, from the indexing of documental resources to their ranking and recommendation.

This paper started with the review of the principal disciplines related to futures and foresight studies and presented a summary of commonly adopted approaches, methodologies, techniques and tools used in related research activities. The paper proposed an architecture for an ICT platform identifying strategic features and services needed to enable the research activity in this domain with ICT tools. It finally propounded a *plastic* integration of platforms with different characteristics and scopes that integrate each other into an ICT ecosystem. Future developments aim at implementing the proposed approach into interoperability features among a set of ICT platforms currently in use.

Acknowledgements. We'd like to thank Maria Angela Biasiotti (ITTIG – CNR, Institute of Legal In-formation Theory and Techniques) who provided insight and expertise that always led to useful, constructive and continuous discussion.

References

1. Accordino, F.: The Futurium—a Foresight Platform for Evidence-Based and Participatory Policymaking. Philosophy & Technology **26**(3), 321–332 (2013)
2. Bell, W.: What do we mean by futures studies?. New Thinking for a New Millennium: The Knowledge Base of Futures Studies **1** (2002)
3. Börjeson, L., Höjer, M., Dreborg, K.-H., Ekvall, T., Finnveden, G.: Scenario types and techniques–towards a user's guide. Futures **38**, 723–739 (2006)
4. Bradfield, R., Wright, G., Burt, G., Cairns, G., van der Heijden, K.: The origins and evolution of scenario techniques in long range business planning. Futures **37**, 795–812 (2005)
5. Brail, R.K., Klosterman, R. (eds.): Planning Support Systems. ESRI Press, Redland (2001)
6. Carlsen, H., Dreborg, K.H., Wikman-Svahn, P.: Tailor-made scenario planning for local adaptation to climate change. Mitigation and Adaptation Strategies for Global Change **18**(8), 1239–1255 (2013)
7. Cassingena, H.J., Cuhls, K., Georghiou, L., Johnston, R.: Future-oriented technology analysis as a driver of strategy and policy (2008)
8. Chermack, T.J.: Studying scenario planning: Theory, research suggestions, and hypotheses. Technological Forecasting and Social Change **72**(1), 59–73 (2005)
9. Conway, M.: Using Foresight to Develop Alternative Visions ATEM Conference, September 2000 (2000)
10. de Jouvenel, B.: Die Kunst der Vorausschau, p. 305. Luchterhand, Berlin (1967)
11. Grzegorz, D., Wm Tompa, F.: Exploring and analyzing documents with OLAP. In: Proceedings of the 5th Ph.D. Workshop on Information and Knowledge. ACM (2012)
12. Durst, C., Durst, M., Kolonko, T., Neef, A., Greif, F.: A holistic approach to strategic foresight: A foresight support system for the German Federal Armed Forces. Technological Forecasting and Social Change (2014)
13. Environmental Protection Agency Willamette Basin Alternaive Futures Anaysis, EPA 600/R-02/045(b) (2002). www.epa.gov/wed/pages/projects/alternativefutures/twopager.pdf
14. Yi, F.,: Entity information management in complex networks. In: Proceedings of the 33rd International ACM SIGIR Conference on Research and Development in Information retrieval. ACM (2010)
15. Ferri, F., Grifoni, P., Caschera, M.C., D'Andrea, A., D'Ulizia, A., Guzzo, T.: An ecosystemic environment for knowledge and services sharing on creative enterprises. In: Proceedings of the 6th International Conference on Management of Emergent Digital EcoSystems, pp. 27–33. ACM, September 2014
16. Bela, G., Beel, J., Hentschel, C.: Scienstein: a research paper recommender system. In: International Conference on Emerging Trends in Computing (2009)
17. Gordon, T., Glenn, J.: Integration, comparisons, and frontiers of futures research methods. In: Futures Research Methodology (Version 2.0). AC/UNU Millennium Project, Washington, DC (2003)
18. Grifoni, P., Ferri, F., D'Andrea, A., Guzzo, T., Praticò, C.: SoN-KInG: a digital eco-system for innovation in professional and business domains. Journal of Systems and Information Technology **16**(1), 77–92 (2014)
19. Haegeman, K., Cagnin, C., Könnölä, T., Collins, D.: Web 2.0 foresight for innovation policy A caseof strategic agenda setting in Europe a ninnovation. Innovation **14**(3), 446–466 (2012)
20. Heger, T., Rohrbeck, R.: Strategic foresight for collaborative exploration of new business fields. Technological Forecasting and Social Change **79**(5), 819–831 (2012)

21. Howe, J.: Crowdsourcing: Why the Power of the Crowd Is Driving the Future of Business. Random House, New York (2006)
22. Kok, K., van Vliet, M., Bärlund, I., Dubel, A., Sendzimir, J.: Combining participative backcasting and exploratory scenario development: experiences from the SCENES project. Technological Forecasting and Social Change **78**(5), 835–851 (2011)
23. Koren, Y., Bell, R.: Advances in collaborative filtering. In: Recommender Systems Handbook, pp. 145–186. Springer US (2011)
24. Palomino, M.A., Vincenti, A., Owen, R.: Optimising Web-based information retrieval methods for horizon scanning. Foresight **15**(3), 159–176 (2013)
25. Soojung-Kim, P.A.: Futures 2.0: rethinking the discipline. Foresight **12**(1), 5–20 (2010)
26. Vu Tran, P.: Social-aware document similarity computation for recommender systems. In: 2011 IEEE Ninth International Conference on Dependable, Autonomic and Secure Computing (DASC). IEEE (2011)
27. Porter, M.E.: Competitive Advantage. Free Press, New York (1985)
28. Noah, R.: Online foresight platforms: Evidence for their impact on scenario planning & strategic foresight. Technological Forecasting and Social Change (2014)
29. Ringland, G.: Scenario Planning: Managingfor the Future. Wiley & Son, London (1998)
30. Rohrbeck, R., Thom, N., Arnold, H.: IT tools for foresight: The integrated insight and response system of Deutsche Telekom Innovation Laboratories. Technological Forecasting and Social Change (2013)
31. Schafer, J.B., Konstan, J.A., Riedl, J.: E-commerce recommendation applications. Data Mining and Knowledge Discovery **5**(1–2), 115–153 (2001)
32. Schatzmann, J., Schäfer, R., Eichelbaum, F.: Foresight 2.0-Definition, overview & evaluation. European Journal of Futures Research **1**(1), 1–15 (2013)
33. Wang, S., Sun, J., Gao, B.J., Ma, J.: VSRank: A Novel Framework for Ranking-Based Collaborative Filtering. ACM Trans. Intell. Syst. Technol. **5**(3), Article 51, July 2014
34. Smith, H.L.: The social forecasting industry. Climatic Change **11**(1–2), 35–60 (1987)
35. Steinitz, C.: Alternative futures for changing landscapes: the upper San Pedro River Basin in Arizona and Sonora. Island press (2003)
36. Su, X., Khoshgoftaar, T.M.: A survey of collaborative filtering techniques. Advances in Artificial Intelligence **4** (2009)
37. van der Heijden, K.: Scenarios: The Art of Strategic Conversation, 2nd edn. Wiley, Chichester (2005)
38. Vander Wal, T.: Folksonomy Coinage and Definition (2005). http://vanderwal.net/folksonomy.html
39. Zhang, D., et al.: Topic modeling for OLAP on multidimensional text databases: topic cube and its applications. Statistical Analysis and Data Mining **2**(5–6), 378–395 (2009)

First International Workshop in Mining the Social Web (SoWeMine 2015)

Sensing Airports' Traffic by Mining Location Sharing Social Services

John Garofalakis[1], Ioannis Georgoulas[1], Andreas Komninos[2],
Periklis Ntentopoulos[1], and Athanasios Plessas[1(✉)]

[1] Computer Engineering and Informatics Department, University of Patras, Patras, Greece
{garofala,georgoulas,ntentopoul,plessas}@ceid.upatras.gr
[2] Computer and Information Sciences, University of Strathclyde, Glasgow, UK
andreas.komninos@strath.ac.uk

Abstract. Location sharing social services are popular among mobile users resulting in a huge social dataset available for researchers to explore. In this paper we consider location sharing social services' APIs endpoints as "social sensors" that provide data revealing real world interactions, although in some cases, the number of recorded social data can be several orders of magnitude lower compared to the number of real world interactions. In the presented work we focus on check-ins at airports performing two experiments: one analyzing check-in data collected exclusively from Foursquare and another collecting additionally check-in data from Facebook. We compare the two popular location sharing social platforms' check-ins and we show that for the case of Foursquare these data can be indicative of the passengers' traffic, while their number is hundreds of times lower than the number of actual traffic observations.

Keywords: Location sharing services · Foursquare · Facebook · Check-ins · Ubiquitous social computing

1 Introduction

During the last decade, the use of online social networking tools and services became widespread, with mobile devices playing an important role as they allow users to connect with others and share information anytime and anywhere. As social networking platforms engage millions of users, there is a boom in the amount of social data that are produced. These data are proving to be representative of real world phenomena and their analysis allows researchers to get the 'big picture' behind social interactions. As an example, consider the work of Bollen, Mao and Zeng [1] who detected public mood from public tweets and were able, under conditions, to predict stock market behavior. Another example is the real-time detection of events, such as earthquakes [2] or sports events [3], based on social interactions.

Moreover, most social networking platforms allow users to geographically characterize the information they share. All modern mobile devices are equipped with GPS sensors, providing social networking applications with the user's location context. For example, Facebook allows users to check-in (declare one's presence at a location) or geo-

© Springer International Publishing Switzerland 2015
F. Daniel and O. Diaz (Eds.): ICWE 2015 Workshops, LNCS 9396, pp. 131–140, 2015.
DOI: 10.1007/978-3-319-24800-4_11

tag their posts, while Twitter and Flickr permit users to geo-tag their tweets and photos respectively, and Foursquare to check-in at places, rate them and leave tips for others. These interactions result in large exploitable datasets conveying rich semantics and describing patterns of human interactions with their environment. However, this kind of social data constitute only a small portion of actual social interactions. In this paper we aim our attention especially at check-ins, i.e. the sharing of a user's location in location based social networks, such as Facebook and Foursquare and the geo-annotation of information about a venue using spatial and temporal context.

More specifically, we focus on airports' check-ins and our purpose is to examine whether check-in data from such platforms are adequate in order to get a representative picture of the airports' real passenger traffic. Airport venues belong to the "Travel & Transport" category, which is found to be the most popular, regarding the number of per venue check-ins [4]. We consider location sharing social services APIs' endpoints as sensors providing social data that offer useful insight about real passenger traffic. In the framework of our research we performed two chronologically distinct experiments. During the first one we collected and analyzed approximately 2.000.000 foursquare check-ins from 10 airports in a period of 13 months, wanting to explore possible correlations between these social data and real passenger traffic. During the second experiment we collected around 360.000 foursquare check-ins, 1.155.000 Facebook check-ins and 100.000 Facebook likes from 7 airports for 6 months, while our purpose was to verify or reject our findings when including other location sharing services.

The rest of this paper is organized as follows: in section 2, related work on this area is covered. In section 3, we briefly present the check-in feature of Foursquare and Facebook, while in section 4 we describe how our dataset was collected. In section 5, we present our analysis and finally, in section 6, we discuss our conclusions and our directions for further work.

2   Related Work

Our work is not the first on capturing and analyzing location sharing social data. Several other researchers have already understood the value of these social interactions and how their study may reveal real world patterns and trends. However, as far as we know, this effort is the first that attempts to exploit check-ins at airports and examine possible correlation with actual passenger traffic.

Several works focus on using location sharing social data to reveal city dynamics. In [5] it was shown that even scarce check-in data are in line with datasets from other sources (e.g. traffic data, air pollution data) and can be used to build a good model of the city's dynamics. Another work on utilizing social media to understand city dynamics is the one presented in [6]. The authors analyzed approximately 18 million Foursquare check-ins (published as tweets) and proposed a methodology that reveals city clusters called "Livehoods", reflecting the dynamic areas that characterize the city. Another effort on using location sharing data for identifying urban neighborhoods and modeling the users' activities is described in [7].

In [8], the authors used Foursquare and Instagram datasets and tried to better understand location related information as an important aspect of urban phenomena. They analyzed these datasets to observe users' movements and activities, popularity of city regions and in general to capture city dynamics. Another work focusing on analyzing urban check-ins is the one presented in [9], where a Foursquare dataset from a German city returned clear patterns separating areas known for different activities, such as nightlife or daily work. In [10], Foursquare check-ins are used in a preliminary effort to sense and analyze the geo-social activity distribution of the cities of London, Paris and New York.

Silva et al. [11] showed a different use of Foursquare check-ins useful for economic purposes and also able to support social and marketing applications. More specifically, the researchers identified cultural boundaries and similarities (such as food and drink habits) across populations by analyzing location sharing data. In [12] check-ins from Foursquare were used to predict the impact of Olympic Games in local businesses. Finally, in [13] the researchers investigated the impact of check-ins on identifying the optimal location for retail shops.

It is apparent that the research community is highly interested in taking advantage of data published in location sharing social networks in order to reveal hidden geo-social patterns. Such works could be beneficial to a wide range of society aspects, including but not limited to marketing, tourism and the financial sector. However, until now, research mainly focuses on understanding urban dynamics, proving useful for urban planning but disregarding potential gains for other sectors.

3   Foursquare and Facebook Check-ins

Foursquare is a popular location sharing service, with over 45 million users as of January 2014 [12], mainly targeting mobile users, as it is available for the most widely adopted mobile operation systems (including Android and iPhone OS). Foursquare users are able to share their location with their friends, in the form of a check-in at a place (venue) from a list of nearby places. Moreover, users are able to leave "Tips" about venues that may prove useful to other users visiting these places.

The approach Foursquare has taken regarding check-ins (allowing users to selectively report their location or not) has changed location from a state (being somewhere) to an action (doing something) [14]. Foursquare users, according to [14], adapt their check-ins to norms of what they perceive as worthwhile check-ins. They tend to check-in at places they find interesting and avoid check-ins at places considered uninteresting. In [15], Lindqvist et al. also examined why people use Foursquare. They categorized it as a primarily social driven application, with much of the motivation coming from the benefit of sharing location check-ins. However, some users are motivated by keeping a record of the places they check-in. At the same study, the researchers found again that users tend to avoid checking in at places that are not considered interesting or where sharing their location would make them feel embarrassed. Cheng et al. [16] and Rost et al. [17] report that airports are among the top places that Foursquare users check-in.

On the other hand, Facebook introduced Facebook Places on 2010, allowing users to check-in via mobile devices and also tag their friends. Location tagging was later expanded as a more general feature, allowing users to share their locations on status updates, photos and wall posts [18]. Similar norms (e.g. increase of self popularity, optimization of self image etc.) as those described earlier exist also about self-presentation via check-ins for Facebook users [18][19].

Foursquare and Facebook offer public APIs, providing information such as the number of check-ins at a place, the number of unique users that have checked-in, the users that are currently present at this place (in Foursquare) etc. However, these check-ins are anonymous and cannot be assigned to the respective users.

4   Data Collection

For our first experiment, whose purpose was to examine possible correlations between Foursquare data and real passenger traffic, we have used the Foursquare venue API in order to collect our dataset that includes the number of check-ins and the number of new users that have checked-in at each airport. A web based script was running every hour, querying Foursquare about the following 10 European airports: Athens International Airport, Heathrow Airport, Paris CDG, Moscow Sheremetyevo Airport, Madrid Barajas, Budapest International Airport, Rome Fiumicino, Frankfurt am Main, Amsterdam Schiphol and Munich International Airport. We selected some of the busiest European airports and we also included in our research a couple of airports with less traffic (e.g. Athens, Budapest). Each airport's API endpoint is seen as a social data sensor, used to sense Foursquare users' social interactions. Each query to the API returns the number of check-ins at that time and the number of individual users that have checked-in. As a result, by subtracting each number with that of the previous hour we were able to compute the check-ins and the new users that have checked-in within this hour and this information was saved in our database. Our research took place for a period of 13 months, from May of 2013 until the end of May 2014, with the exception of Budapest Airport that was added later (gathered data for this airport expand from July 2013 to May 2014). In total, there were 2.032.273 check-ins and the new users having checked-in during that period were 757.975.

Our analysis, presented in the next section was performed in a monthly base, as we were able to find only monthly statistics for these airports, to compare with. More specifically, we used the statistics published at Anna Aero[1], which is a website dedicated to airline and airports network news and analysis.

In table 1, we summarize our findings for each airport:

For our second experiment, where we wanted to confirm our findings using other location sharing services, we expanded our software incorporating the Facebook API and as a result we were able to collect data from both location sharing social services. At first, we had to identify the Facebook pages that correspond to the airport venues from Foursquare that were included in our first experiment. In this case, our research took place for a period of 6 months, from October of 2014 until the end of March 2015. During this period the Facebook pages for Rome Fiumicino and Paris CDG

---

[1]   http://www.anna.aero

airports that we queried the API for were merged with other existing pages, however we failed to notice this fact. As a result, we were unable to capture check-ins from the merged pages. Moreover, for an unknown reason, the Facebook API returned an abnormally low number of new check-ins for Amsterdam Schiphol airport. Consequently, we were forced to exclude these three airports from our analysis. As shown in table 2, we were left with seven airports, recording in total 357.656 Foursquare check-ins, 86.625 new Foursquare users having checked-in, 1.155.395 Facebook check-ins and 97.454 Facebook likes.

Table 1. Summary of Foursquare airport data from first experiment

Airport	# Check-ins	# New Users	# Passengers
Athens Intl	96.100	26.635	14.255.499
Heathrow	358.155	131.142	79.040.930
Frankfurt am Main	188.346	75.040	63.636.892
Paris CDG	212.289	105.398	68.393.212
Amsterdam Schiphol	236.206	80.464	58.288.233
Madrid Barajas	153.329	61.948	43.569.777
Munich Intl	161.406	57.773	42.198.301
Moscow SVO	480.295	156.211	32.792.835
Rome Fiumicino	85.007	41.153	39.836.419
Budapest Intl	61.140	22.211	7.922.530
Total	**2.032.273**	**757.975**	**449.934.628**

Table 2. Summary of Foursquare and Facebook airport data from second experiment

Airport	# 4sq Check-ins	# 4sq New Users	# FB Check-ins	# FB Likes	# Passengers
Athens Intl	34.624	6.263	153.673	7.928	5.390.275
Heathrow	77.636	20.640	348.465	37.228	28.122.242
Frankfurt am Main	45.447	15.073	286.256	26.958	21.832.189
Madrid Barajas	34.359	10.980	108.456	2.920	16.516.643
Munich Intl	39.877	11.325	125.521	11.322	14.984.963
Moscow SVO	102.961	15.130	67.130	3.543	11.283.720
Budapest Intl	22.752	72.14	65.894	7.555	4.180.418
Total	**357.656**	**86.625**	**1.155.395**	**97.454**	**102.310.450**

We should note here that at the time of our analysis there were not actual passenger data for all airports (with the exception of the Budapest airport) for March 2015.

5   Data Analysis

In this section we present the results of our data analysis. As already mentioned, in our first experiment our purpose was to examine whether there is a correlation between social data from Foursquare and the actual number of passengers travelled in each airport. Figure 1 depicts for each airport a graph showing the actual number of passengers (orange line), the number of check-ins (blue line) and the number of new users having checked-in (grey line) on a monthly base. The left vertical axis corresponds to the number of passengers (orange line), while the right vertical axis corresponds to the social data gathered (blue and grey lines).

A statistical analysis of the monthly observations for all airports reveals a strong positive correlation between the number of check-ins and the actual number of passengers, which is statistically significant (r_s=0.588 p<0.01). Moreover, there is an even stronger positive correlation between the number of new users having checked-in and the number of passengers, which is also statistically significant (r_s=0.666 p<0.01). Finally, a strong statistically significant positive correlation was also found between the number of new users and the number of check-ins (r_s=0.956 p<0.01). These correlations confirm the similar trends that follow the graphical representations of these variables.

When examining data from each airport separately, results are quite different. Statistically significant correlations between the number of check-ins and the number of passengers were found for the following airports: Athens (r_s=0.736 p<0.01), Heathrow (r_s=0.698 p<0.01), Frankfurt (r_s=0.648 p<0.05), Paris (r_s=0.725 p<0.01), while for the airport of Amsterdam the correlation is not marginally statistically significant (r_s=0.523 p=0.066). However, statistically significant correlation is found between the number of new users and the number of passengers when analyzing each airport data, for all cases except for the airport of Munich: Athens (r=0.900 p<0.01), Heathrow (r=0.883 p<0.01), Frankfurt (r=0.935 p<0.01), Paris (r_s=0,841 p<0.01), Amsterdam (r=0.605 p<0.05), Madrid (r=0.563 p<0.01), Moscow (r=0.929 p<0.01), Rome (r=0.576 p<0.05) and Budapest (r=0.638 p<0.05). Again a positive statistically significant correlation was found between the number of new users and the number of check-ins for each airport, except for the international airport of Moscow.

Having found that social data from Foursquare are statistically correlated to actual passenger traffic, we wanted to check if this is also true for other location sharing social networking services. Working with the recorded data from the second experiment we were able to statistically analyze Facebook data in relation to Foursquare data and real-world observations. An important finding of this analysis was that there is no statistically significant correlation between Foursquare check-ins when compared to Facebook check-ins or Facebook likes.

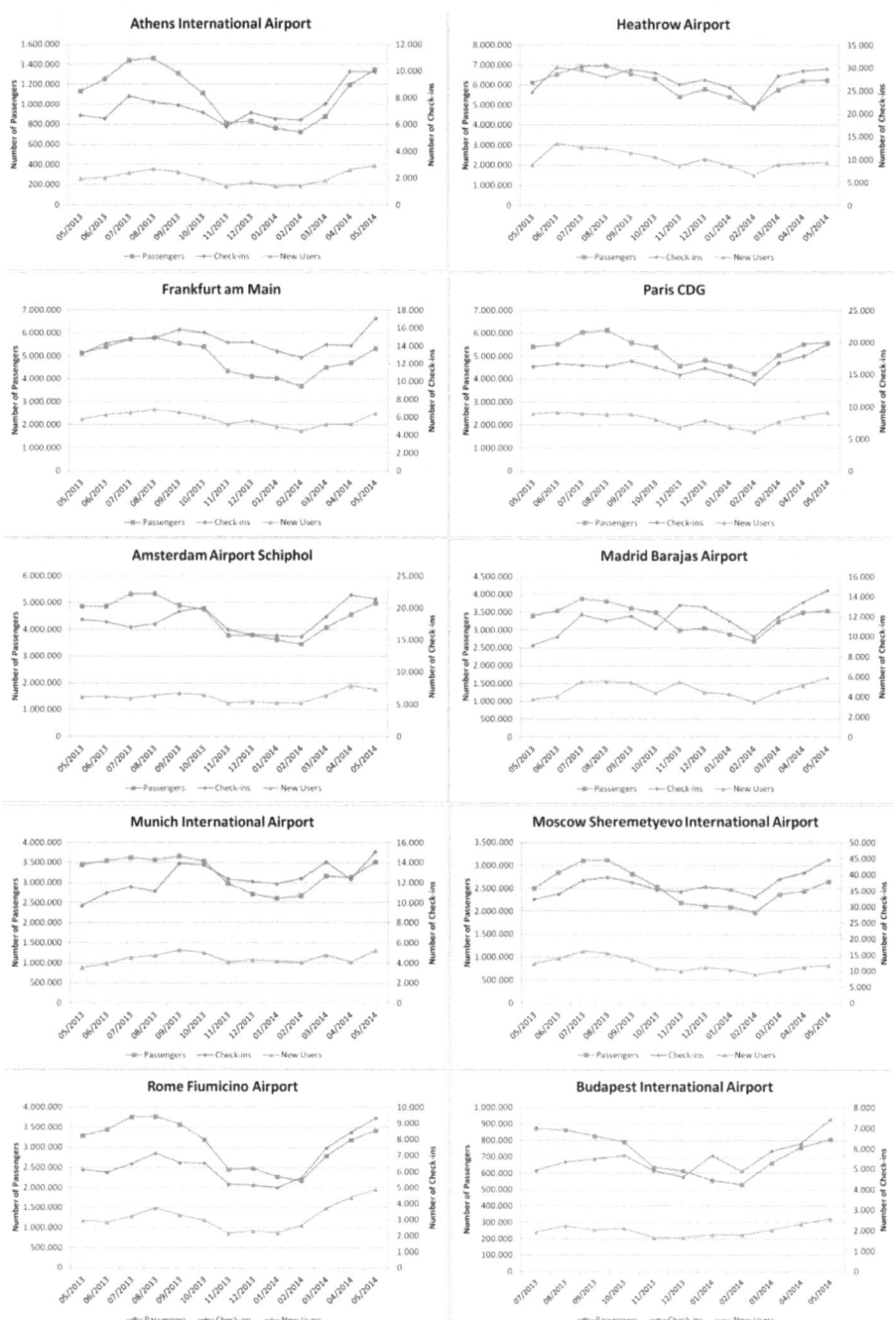

Fig. 1. Data graph for each airport in first experiment

Moreover, the statistical analysis of Facebook data from each airport venue does not reveal statistically significant correlation between Facebook check-ins or likes and real world observations, with the exception of Facebook likes for the Moscow Sheremetyevo airport ($r=0.978$ $p<0.05$). Finally, once again we observe that new Foursquare users having checked in at the airport venues is the best indicator for actual passengers traffic, since we found strong statistically significant correlations for the following airports: Athens ($r=0.928$ $p<0.05$), Heathrow ($r=0.952$ $p<0.05$), Madrid ($r=0.956$ $p<0.05$), Moscow ($r=0.964$ $p<0.01$), while for Frankfurt ($r=0.832$ $p=0.08$) and Munich ($r=0.870$ $p=0.055$) the correlations are not only marginally statistically significant.

6   Discussion and Future Work

In this paper we considered location sharing social networks as sensors providing geo-social data, which we analyzed to reveal correlations with real-world data. We focused on airports, as airport traffic is an important financial factor for national economies and is often used as a tourism indicator [20]. Moreover, as mentioned earlier, venues belonging to the Tourism & Travel category such as airports are the most popular places where location sharing social services' users check-in and as a result we considered them as an interesting starting point for our research. While the analysis of check-ins is popular (as presented in the related work section), as far as we know this is the first study focusing exclusively on airport venues.

When examining total observations from all sensors – airports in our first experiment we found that the number of recorded Foursquare check-ins and even more the number of new users performing a check-in are strongly correlated to the actual number of passengers that traveled to and from those airports. Thus, location sharing data seem to be representative of real world data for the airport venue type, which is encouraging for the adoption of geo-social data analysis for other domains apart from urban dynamics, which is mostly referenced in related literature. Consequently, interested parties are able to capture the traffic trend for airports before official data are announced. In addition to this, one could argue that an idea worth to study is this of incorporating airport check-ins into airport traffic prediction models, which is a subject that has attracted considerable attention from the research community [21].

However, a significant finding from our second experiment was that we were not led to the same conclusion when examining Facebook data. As a result, it seems that not all location sharing social networks generate geo-social data indicative of the real-world situations, at least for the case of airport venues. While we report this finding with the reservation of the small number of airports and the short duration of data collection period for our second experiment, we believe that it is an important contribution, since as far as we know it is the first research effort on comparing geo-social data referencing the same places and originating from different social services. Such a comparison of different users' check-ins patterns and attitudes in different location sharing social services is a challenge that we intend to undertake in the future in large-scale experiments.

When analyzing data from each airport separately in our first experiment, we found a correlation between the number of check-ins and airport traffic for 5 out of 10 airports. On the other hand, the analysis showed a correlation between new users and passengers for 9 out of 10 airports. This observation is in line with the graphs in Figure 1, where one can see that the line corresponding to new users approximates more closely the line of airport traffic than that of check-ins. It seems, then, that the number of new users checking-in through Foursquare is a better indicator of airport traffic. This is quite important, in our opinion, since these numbers are hundreds of times lower than the recorded numbers of passengers.

Moreover, another observation from both experiments is that while the extracted Foursquare social data are representative of the general trend regarding airports' traffic, it seems that we cannot use them to directly compare traffic between airports. This observation confirms Rost et al. [17] findings, according to which rankings of airport venues in terms of check-ins does not keep up with rankings in terms of passengers. We confirm that apart from check-ins, this is also true for new users having checked-in at airports. For example, one can see in table 1 that in Moscow airport there were much more check-ins than in Heathrow, but less than half the passengers, while another such case is the comparison between Frankfurt and Amsterdam airports. A possible explanation for this remark could be based on the different location sharing culture of the user types visiting these airports (e.g. youth vs elder or business users). In addition to this, each airport follows a different social media marketing campaign, which may lead to a different user engagement type [22].

In the future, we intend to extend our research to include more airports, in order to ensure that our remarks apply for airports worldwide and are not limited to the European territory. Moreover, we intend to add more venue types in our analysis (e.g. train stations, ports etc.) in order to examine if social data are also representative of passenger traffic for these cases. Finally, we are also currently putting research effort to investigate possible correlation between weather conditions and check-ins in venues such as parks, beaches etc.

References

1. Bollen, J., Mao, H., Zeng, X.: Twitter Mood Predicts the Stock Market. Springer Journal of Computational Science **2**(1), 1–8 (2011)
2. Sakaki, T., Okazaki, M., Matsuo, Y.: Earthquake shakes twitter users: real-time event detection by social sensors. In: Proceedings of the 19th International Conference on World Wide Web (WWW 2010), pp. 851–860. ACM, New York (2010)
3. Chakrabarti, D., Punera, K.: Event summarization using tweets. In: Proceedings of the Fifth International Conference on Weblogs and Social Media (ICWSM 2011), pp. 66–73 (2011)
4. Li, Y., Steiner, M., Wang, L., Zhang, Z.L., Bao, J.: Exploring venue popularity in foursquare. In: Proceedings of IEEE INFOCOM Workshops (2013)
5. Komninos, A., Stefanis, V., Plessas, A., Besharat, J.: Capturing Urban Dynamics with Scarce Check-In Data. IEEE Pervasive Computing **12**(4), 20–28 (2013)
6. Cranshaw, J., Schwartz, R., Hong, J., Sadeh, N.: The livehoods project: utilizing social media to understand the dynamics of a city. In: Proceedings of the Sixth International Conference on Weblogs and Social Media (ICWSM 2012), pp. 58–65 (2012)

7. Zhang, A., Noulas, A.,Scellato, S., Mascolo, C.: Hoodsquare: modeling and recommending neighborhoods in location-based social networks. In: Proceedings of the 2013 International Conference on Social Computing (SOCIALCOM 2013). IEEE Computer Society, pp. 69–74 (2013)
8. Silva, T., Vaz de Melo, P., Almeida, J., Salles, J., Loureiro, A.: A comparison of foursquare and instagram to the study of city dynamics and urban social behavior. In: Proceedings of the 2nd ACM SIGKDD International Workshop on Urban Computing (UrbComp 2013), Article No 4. ACM, New York (2013)
9. Rösler, R., Liebig, T.: Using data from location based social networks for urban activity clustering. In: Geographic Information Science at the Heart of Europe. Lecture Notes in Geoinformation and Cartography, pp. 55–72. Springer (2013)
10. Phithakkitnukoon, S., Olivier, P.: Sensing urban social geography. using online social networking data. In: Proceedings of the Fifth International Conference on Weblogs and Social Media (ICWSM 2011), pp. 36–39 (2011)
11. Silva, T., Vaz de Melo, P., Almeida J., Musolesi, M., Loureiro, A.: You are what you eat (and drink): identifying cultural boundaries by analyzing food & drink habits in foursquare. In: Proceedings of 8th AAAI Intl. Conf. on Weblogs and Social Media (ICWSM 2014) (2014)
12. Georgiev, P., Noulas, A., Mascolo, C.: Where businesses thrive: predicting the impact of the olympic games on local retailers through location-based services data. In: Proceedings of 8th AAAI Intl. Conf. on Weblogs and Social Media (ICWSM 2014) (2014)
13. Karamshuk, D., Noulas, A., Scellato, S., Nicosia, V., Mascolo, C.: Geo-spotting: mining online location-based services for optimal retail store placement. In: Proceedings of the 19th ACM SIGKDD International Conference on Knowledge Discovery and Data Mining (KDD 2013), pp. 793–801. ACM, New York (2013)
14. Cramer, H., Rost, M., Holmquist, L.E.: Performing a check-in: emerging practices, norms and 'conflicts' in location-sharing using foursquare. In: Proceedings of the 13th International Conference on Human Computer Interaction with Mobile Devices and Services (MobileHCI 2011), pp. 57–66. ACM, New York (2011)
15. Lindqvist, J., Cranshaw, J., Wiese, J., Hong, J., Zimmerman, J.: I'm the mayor of my house: examining why people use foursquare - a social-driven location sharing application. In: Proceedings of the SIGCHI Conference on Human Factors in Computing Systems (CHI 2011), pp. 2409–2418. ACM, New York (2011)
16. Cheng, Z., Caverlee, J., Lee, K., Sui, D.: Exploring millions of footprints in location sharing services. In: Proceedings of the Fifth International Conference on Weblogs and Social Media (ICWSM 2011), pp. 81–88 (2011)
17. Rost, M., Barkhuus, L., Cramer, H., Brown, B.: Representation and communication: challenges in interpreting large social media datasets. In: Proceedings of the 2013 conference on Computer supported cooperative work (CSCW 2013), pp. 357–362. ACM, New York (2013)
18. Wang, S., Stefanone, M.: Showing Off? Human Mobility and the Interplay of Traits, Self-Disclosure, and Facebook Check-Ins. Social Science Computer Review 31(4), 437–457 (2013)
19. Wang, S.: "I Share, Therefore I Am": Personality Traits, Life Satisfaction, and Facebook Check-Ins. Cyberpsychology, Behavior, and Social Networking 16(12), 870–877 (2013)
20. Halpern, N.: Lapland's Airports: Facilitating the Development of International Tourism in a Peripheral Region. Scandinavian Journal of Hospitality and Tourism 8(1), 25–47 (2008). Taylor & Francis
21. Wang, M., Song, H.: Air Travel Demand Studies: A Review. Journal of China Tourism Research 6(1), 29–49 (2010)
22. Vanauken, K.: Using social media to improve customer engagement and promote products and services. Journal of Airport Management 9(1), 109–117 (2015)

An Approach for Mining Social Patterns in the Conceptual Schema of CMS-Based Web Applications

Vassiliki Gkantouna[1(✉)], Tsakalidis Athanasios[1], Giannis Tzimas[2], and Emmanouil Viennas[1]

[1] Department of Computer Engineering and Informatics, University of Patras,
Rio Patras 26500, Greece
gkantoun@ceid.upatras.gr
[2] Department of Computer and Informatics Engineering, Technological Educational
Institute of Western Greece, M. Alexandrou 1, Koukouli Patras 26334, Greece

Abstract. The advent of the emerging social technologies has transformed the Web to a place where users can turn for social interaction, content consumption and opinion making. As social networks are becoming more ubiquitous, they set new requirements to the needs of modern enterprises which now need web applications that can incorporate social networking features. This fact leads to the need for tools supporting developers during the design and development of such socially enabled applications. In this work, we focus on CMS-based web applications which exploit social networking features and propose a model-driven approach to evaluate their hypertext schema in terms of the incorporated design fragments that perform a social network related functionality. We have developed a methodology which based on the identification and evaluation of design reuse within an application's hypertext schema detects a set of recurrent design solutions (i.e. configurations of hypertext elements) denoting either design inconsistencies or effective reusable social design structures that can be used as building blocks for implementing certain social behavior in future designs.

Keywords: Design pattern · Design evaluation · CMS · Web application · Hypertext schema · Social networking features

1 Introduction

Modern web applications support a variety of sophisticated functionalities incorporating advanced business logic, one-to-one personalization features and multimodal content delivery. Nowadays, with the advent of the emerging social technologies, they also incorporate social networking features such as social media integration, commenting and sharing in order to foster social interaction. At the same time, their increasing complexity makes it hard for developers to build software products with quality design. In the category of social network enabled web applications, enterprises need solutions that can incorporate a peculiar subset of the large set of features supported by the

© Springer International Publishing Switzerland 2015
F. Daniel and O. Diaz (Eds.): ICWE 2015 Workshops, LNCS 9396, pp. 141–152, 2015.
DOI: 10.1007/978-3-319-24800-4_12

various social platforms, which properly comply with the specific context of the company. This raises the problem of developing applications which integrate such a heterogeneous set of social networking services into a single application. Furthermore, due to the lack of time, in most cases developers often reuse their previous work without properly adjusting it to meet the ad-hoc needs of the new project at hand. This leads to cases of bad reuse and regarding the social features in particular, it happens that they are used in ways which are not appropriate for achieving certain social communication effects. These reasons highlight the need for tools that allow developers inspect the design of social applications focusing specifically on the utilization of the social features in their design model, even at the conceptual level.

Having in mind the aforementioned need, we propose a model-driven approach for evaluating the use of social networking features in the design of web applications, aiming to identify reusable design structures in their conceptual schema which are used by the designers to implement a certain social functionality, paving the way for social design patterns. To detect such structures, we have developed a methodology to perform a pattern-based analysis and evaluation of an application's conceptual schema which is specified in the WebML [1] notation which comprises a set of modeling concepts encapsulating the logic of the interaction with the social platforms [2]. The proposed approach is guided by the work presented in [3] for evaluating the hypertext schema of web applications under the viewpoint of design reuse. However, the methodology in [3] cannot be applied to any web application, since it does not provide a practical way to produce the application's hypertext schema which is a key point of the methodology. To overcome this issue, we have narrowed down the methodology's scope to the domain of Content Management Systems (CMSs) for two main reasons. First, CMSs provide a common base of source code that can be systematically processed in order to extract the hypertext schema of an application in an automatic manner. Secondly, we believe it is important to study the dynamics of social networking features in the CMS domain, due to the increasing popularity of CMS platforms and considering the shift we have witnessed in the last decades to content management.

First, we extract the hypertext schema of a CMS-based web application. Then, we detect all the incorporated design fragments (i.e. configurations of WebML hypertext elements) within the recovered schema in which WebML social units participate. With the objective to mine 'social' design patterns, we focus on those fragments which are repeated throughout the hypertext schema, possibly deriving from design reuse, on the grounds that the recurrence can indicate a pattern for implementing a certain social task. To capture such fragments, we map the problem to the graph theory domain and utilize a graph mining algorithm to perform a pattern-based analysis of the hypertext schema. This results in detecting not only the recurrent design fragments but also their variants which extend the core configuration of WebML elements composing each fragment with all the alternatives in which it can start or terminate. Finally, in order to examine whether the identified fragments can form the basis for possible social design patterns, we evaluate their appropriateness by applying a number of metrics related to their implications on the overall application schema's quality.

By applying the methodology on a CMS-based web application, developers can have access to a set of recurrent design structures within the application's hypertext schema which based on their evaluation indicate two main facts. On one side, they can possibly indicate effective reusable design solutions that can be used as building blocks for implementing certain social behavior in future designs, facilitating the discovery of social design patterns in the CMS domain. On the other hand, they can indicate cases of inconsistent use of social features implying the need for refactoring. The remaining of this paper is organized as follows: Section 2 provides an overview of the related work. Section 3 presents in detail the methodology for mining reusable design solutions in the conceptual schema of CMS-based applications, Section 4 describes an exemplifying paradigm for demonstrating the methodology, while Section 5 concludes the paper and discusses future steps.

2 Related Work

In this paper we propose a model-driven approach for investigating the role of social networks in the design of the specific domain of CMS-based web applications. Several studies can be found in the literature addressing the use of social technologies in the design of web applications. Among them, we can mention the work [2] in which the authors describe a model-driven approach specifically focused on the development of web applications that exploit social features. They describe an extension of the WebML comprising a set of modeling concepts, namely social units that encapsulate the logic of the interaction with various social platforms. Based on these units, they identify a set of design patterns for solving the most common requirements for socially enabled applications. In another work [4], authors have analyzed a number of top-rank Web 2.0 community applications and distilled a number of recurring design patterns specified in WebML notation. The analysis of the pattern set lead to the individuation of core concepts that are the main focus of the social activities in Web 2.0 applications. Another approach based on the requirements engineering paradigm for the development of customized social network applications can be found in [5]. All the aforementioned approaches refer to design patterns which are derived from typical usage cases in the design of socially enabled web applications. Such patterns are devised by experienced designers who review a set of successful social applications and define a number of design templates to solve common design problems. The key difference of our approach from other similar approaches is that we provide a methodology accompanied by a toolset for supporting the automated identification of the reusable social patterns which can possibly lead to social design pattern. In other words, we attempt to automate the design patterns discovery process. By applying the methodology to a website, all the recurrent social design structures within its hypertext schema are available to the designers. This way, designers can have the opportunity to inspect them and examine the possibility that they can be used as building blocks characterizing a set of social functionalities in the future projects.

3   A Methodology for Mining Reusable Social Design Solutions

In this section, we present the methodology for detecting the recurrent design fragments deriving from potential design reuse within the hypertext schema of existing
CMS-based web applications. In order to demonstrate the potential of the proposed
methodology, we have examined the Joomla! [6] CMS platform and we have developed a toolset (available at http:150.140.142.61:8282/CMSModeling) to support the
application of the methodology on websites which are built using Joomla!. In the
context of this work, we have focused on a set of typical social features which are
provided by the most popular Joomla! extensions for the social web (Table 1). Among
others, these features provide support for:

- Social network authentication via user's social media account.
- Social sharing buttons for providing the ability to users to share and comment web-
  site's content to Facebook, Twitter and LinkedIn.
- Social posting for allowing users push website's content to their Facebook Page.
- Facebook invitations allowing users to invite their Facebook friends to the website.
- Social media stream.
- Creation of a custom social network supporting a community.

The methodology comprises three distinct phases. In the first phase (section 3.1),
we have developed a process (Figure 1) which given the URL of a website, it performs an analysis of its HTML pages with the purpose to reverse engineer them to
extract the website's hypertext schema. In the second phase (section 3.2), in order to
detect the recurrent social design fragments lying in the extracted hypertext schema,
we transform it into a directed graph and then we apply a graph mining algorithm on
the graph. This results in identifying a set of recurrent design structures specified as
configurations of WebML hypertext elements. Among these structures, we select the
ones that include WebML social units considering the fact that they can possibly implement a social network related task. In the third phase (section 3.3), we apply a
number of evaluation metrics on the identified design fragments in order to categorize
them towards the appropriateness of reuse, facilitating the designers to inspect them
and gain a better understanding of the application's design.

3.1   Hypertext Schema Extraction Process

The goal of the hypertext schema is to specify the organization of the front-end interfaces of a web application. Thus, in order to extract the hypertext schema of a given
website, we have developed a process which by reverse engineering the source code
of the website's HTML pages, recovers the organization of these pages in terms of the
structural and navigational design elements that compose their hypertext and then
produces their representation in WebML notation. In the context of Joomla!, such
elements are the components and modules which specify the layout of an HTML
page. Intuitively, components determine the layout of the main part of the page and
modules determine the layout of the front-end elements lying in the peripheral parts of
the page. The process includes three main steps as depicted in Figure 1.

Having as a starting point the URL of a website, we initially employ a web crawler (supported by the Web Crawler tool) which visit the website's pages and stores them locally for subsequent analysis. Then, we parse the source code of these pages in order to identify the various configurations of components and modules that specifies each page's layout. For their identification, we have analyzed a plethora of Joomla! websites and we have registered a set of characteristic values (the complete set is available at http:150.140.142.61:8282/CMSModeling) of the class attribute occurring in the HTML tags of a web page which identifies the components and modules. Regarding the social features supported by Joomla!, we have included the features provided by the most popular Joomla! extensions for social networking.

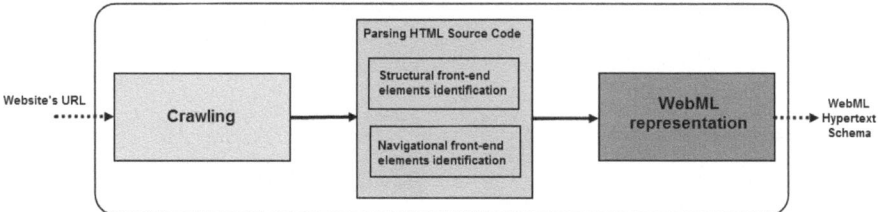

Fig. 1. Hypertext Schema Extraction Process

An indicative list (the complete list is available at http:150.140.142.61:8282/CMS-Modeling) of these extensions is available in Table 1.

Table 1. An indicative list of popular Joomla! extensions for social networking

Feature	Joomla! extension
Community creation	• Community Builder • JomSocial • EasySocial
Social media integration (Facebook, Twitter, Youtube, Vimeo, Instagram, Flickr, LinkedIn, Pinterest, Google+, rss, reddit, delicious)	• JA Social Feed Plug-in • Facebook Wall Feeds • Article-Generator • AutoTweet NG • Youtube Gallery

By parsing the code of every HTML page, we detect all the occurrences of those characteristic values and thus we manage to recover the page's organization in terms of components and modules. We also parse the HTML DOM tree of each page to extract additional information about the identified components and modules, such as the content they display and their exact layout format, in case that they can have various layout options. For example, when we identify a menu, we compute its depth in order to determine whether it should be represented as an index unit or as a hierarchical index unit. Next, the representation of the HTML pages in WebML notation follows in order to generate the website's WebML hypertext schema in the form of an XML file containing the textual syntax of those representations. We have defined appropriate WebML representations as compositions of WebML units for all the components, modules and extensions involved in this work. An example can be found in Figure 2 which depicts the WebML representation of a blog featured page. The hypertext schema is defined as a site view representing the entire website which contains a set of WebML pages, each one corresponding to the website's HTML pages. Every WebML page is then assembled by producing the WebML representations of the previously identified component and modules which compose the corresponding HTML page. The task of identifying the pages' organization as well as their representation in WebML is supported by the 'Hypertext Schema Extractor' tool which produces an XML file containing the website's hypertext schema.

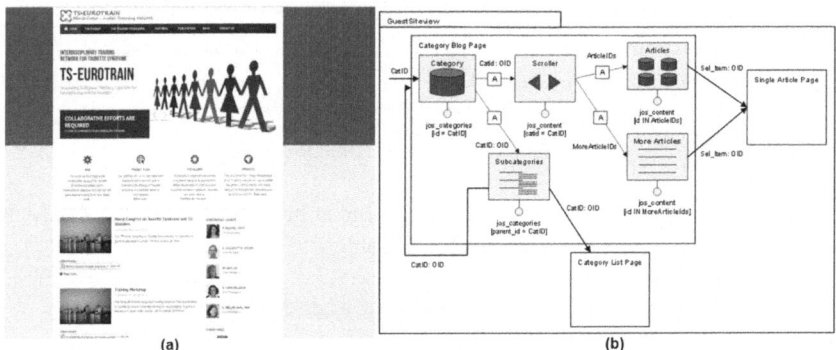

Fig. 2. The WebML representation of blog featured page.

3.2   Mining Reusable Design Solutions

In order to detect the recurrent social design fragments lying within the hypertext schema, we reduce the problem of their identification into the subgraph isomorphism problem which in its general form is synopsized to finding whether the isomorphic image of a subgraph exists in a larger graph. The latter problem has proven to be NP-complete. However, quite a few heuristics have been proposed to face this problem among which we have adopted the approach of gSpan [7].

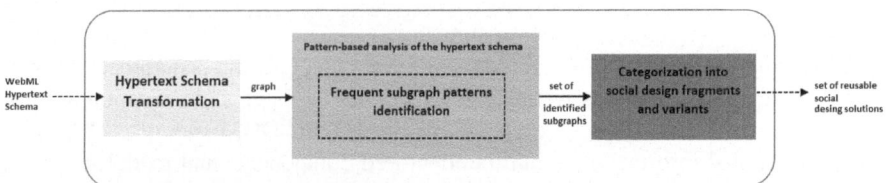

Fig. 3. The process of mining reusable design solutions

To apply the gSpan algorithm on the hypertext schema, we transform it into a directed graph (section 3.2.1). Then, the application of gSpan on the graph returns a set of the recurrent subgraphs occurring within the graph (section 3.2.2). These subgraphs represent the design fragments i.e., the configurations of WebML hypertext elements, which are recurrent within the hypertext schema (Figure 3). Due to the fact that that in this work we are focused specifically in mining social patterns, we select only the subgraphs which represent a design fragment that include at least one WebML social unit.

3.2.1   Hypertext Schema Transformation

We define the hypertext schema's site view as a directed graph of the form G (V, E, f_V, f_E) comprising a set of nodes V, a set of edges E, a node-labeling function f_V: V → Σ_V and an edge-labeling function f_E: E → Σ_E. The function f_V assigns labels from the alphabet Σ_V = {Siteview, Page, DataUnit, MultidataUnit, IndexUnit, HierarchicalIndexUnit, MultichoiceIndexUnit, EntryUnit, ScrollerUnit} to the nodes in V, representing the different types of WebML hypertext elements. Similarly, function f_E assign labels from the alphabet Σ_E = {S_P, P_U, U_P, U_U} to the edges in E. The label S_P denotes the containment of a page in a site view, the label P_U denotes the containment of a content unit within a page, the label U_P denotes the link from a content unit to a page and finally the label U_U denotes the link between content units. To transform the hypertext schema produced in section 3.1 into a directed graph, we parse its XML textual representation and produce its equivalent graph representation. The transformation procedure is supported by the 'Hypertext Graph' tool which produces the graph in the form of a text file in order to provide it as an input to the gSpan algorithm.

3.2.2   Mining the Social Recurrent Design Fragments

As we mentioned earlier, the application of gSpan on the graph created in the previous step returns a set of the recurrent design fragments. Intuitively, gSpan traverses the graph and identifies all the occurrences of the frequent substructure patterns i.e., subgraphs within the graph, performing in this way a pattern-based analysis of the hypertext schema. By detecting the occurrences of these subgraphs in the graph which basically represent configurations of WebML hypertext elements, we manage to locate the recurrent design structures. This task is performed by using the Parsemis tool [9] which returns a text file containing the set of the identified subgraphs occurrences.

Among the identified subgraphs, there are cases of structures that have in common a part which is composed by an identical configuration of WebML elements. In these cases, we consider as a design fragment the core configuration of WebML elements which is identical in all the structures and we also consider as the design fragment's variants the structures which extend the core configuration of WebML elements with all the alternatives in which it can start or terminate. By parsing the text file created above, we separate the identified structures into design fragments and their variants. Among these structures, we select the ones that include WebML social units considering the fact that they can possibly implement a social networking related task and store them in a repository along with the number of their occurrences in the entire graph.

In order to determine whether the occurrences of the identified design fragments and/or their variants derive from developer's attempt to perform design reuse for accomplishing a certain purpose, apart from the recurrence at the hypertext level, we additionally have to examine the recurrence in the organization of the content displayed by the WebML elements that form the core configuration of a design fragment. Assuming we cannot have access to the underlying data schema of a website, we attempt to approximately capture this organization based on the semantic similarity measurement of the content published by the Joomla! design elements that correspond to the WebML elements of the core configuration. This is achieved by using a WordNet-based semantic similarity algorithm which given the contents published by two corresponding WebML elements, computes a semantic similarity score determining how similar their meanings are. Due to the large number of the identified occurrences, we consider as reusable design solutions solely the ones having an average semantic similarity score above the threshold of 0.5.

3.3   Evaluating the Appropriateness of Reuse

In order to evaluate the identified recurrent social design solutions towards the appropriateness of their reuse, we apply a set of metrics on them. The evaluation is twofold: first, we evaluate the consistency of the design fragments based on the consistent use of their variants throughout the website's hypertext schema (section 3.3.1). Secondly, we quantify the impact of each solution to the hypertext schema's quality by taking into consideration a number of evaluation factors characterizing each individual fragment/variant (section 3.3.2).

3.3.1   Consistent use of the Reusable Design Solutions

Fraternali et al. [8] have introduced a methodology for the evaluation of the consistent application of predefined WebML design patterns within the conceptual schema of an application. We utilize and extend this methodology by introducing metrics depicting the consistent application of a design fragment's variants throughout the application' hypertext schema. To achieve that, we introduce two metrics that represent the statistical variance of the occurrence of N termination or starting variants of the identified fragments, normalized according to the best case variance. These metrics are called Start-Point Metric (SPM) and End-Point Metric (EPM) respectively. SPM is defined as (EPM is defined in an analogous way):

$$SPM = \sigma^2 / \sigma_{BC}^2 \tag{1}$$

In (1) σ^2 is the statistical variance of the starting variants occurrences calculated according to the formula (2), where N is the number of variants of the current pattern addressed by the metric, while p_i is the percentage of occurrences for the i-th configuration variant.

$$\sigma^2 = \frac{1}{N} \sum_{i=0}^{N} \left(p_i - \frac{1}{N} \right)^2 \tag{2}$$

σ^2_{BC} is the best case variance and it is calculated by (2), assuming that only one variant has been coherently used throughout the application. The last step in the metrics definition is the creation of a measurement scale, which defines a mapping between the numerical results obtained through the calculus method and a set of meaningful and discrete values. According to the scale types defined in the measurement theory, the SPM metric adopts an ordinal nominal scale; each nominal value in the scale expresses a consistency level, corresponding to a range of numerical values of the metrics as defined in Table 2. The same scale covers the EPM metric as well.

Table 2. The SPM metric measurement scale

Metric's range	Measurement scale value
$0 \leq SPM < 0.2$	Insufficient
$0.2 \leq SPM < 0.4$	Weak
$0.4 \leq SPM < 0.6$	Discrete
$0.6 \leq SPM < 0.8$	Good
$0.8 \leq SPM \leq 1$	Optimum

For every design fragment, we classify the occurrences of its variants in two distinct categories based on its corresponding metric value. We create a first category including all the occurrences of the variants having low consistency level ($0<SPM/EPM<0.6$) and a second category including the ones having high consistency level ($0.6<SPM/EPM<1$). We consider the variants of the first category as cases of bad reuse implying the need for refactoring, whereas we consider the variants of the second category as indication of effective reusable design solutions that can lead to the discovery of new design patterns.

3.3.2 Self-Evaluation of the Reusable Design Solutions

In order to provide a more accurate evaluation of the appropriateness of reuse, we also define a number of metrics based on the particular features of the fragments/variants. The first metric is the frequency (f) representing the number of the occurrences of a fragment/variant. It is comprised by: a) f_o: the overall number of the fragment's/ variant's occurrences and b) f_p: the number of the distinct pages that the fragment/variant occurs. For a given fragment/variant j, we compute its frequency f_j as follows:

$$f_j = \frac{f_{o,j}}{f_{o,max}} + \frac{f_{p,j}}{N_p} \qquad (3)$$

$f_{o,\,max}$ is the maximum value of the corresponding metric (i.e. the max number of occurrences of a fragments/variant) and N_p is the number of the pages that the entire hypertext schema contains.

The length of a variant in terms of the number of the WebML elements it contains is another important factor. Intuitively, the repetition of a larger pattern cannot be a random event, but rather the result of an intended design solution. The larger the length of a pattern is, the higher is the possibility to detect an effective reusable solution. Thus, we define the length (l) that denotes the number of WebML elements involved in a fragment/variant. For a given fragment/variant j, we compute its length l_j as:

$$l_j = \left(\frac{l_{u,j}}{l_{u,max}} + \frac{l_{p,j}}{l_{p,max}} \right) \times \frac{N_{main,j}}{N_{peripheral,j}} \qquad (4)$$

l_u is the number of the WebML content units that are involved in the fragment/variant, $l_{u,max}$ is the maximum value of the corresponding metric (i.e. the max number of content units in a fragment/variant), l_p is the number of the WebML pages that are involved in the fragment/variant, $l_{p,max}$ is the maximum value of the corresponding metric (i.e. the max number of pages in a fragment/variant), N_{main} is the number of the components that the variant contains and $N_{peripheral}$ is the number of the modules.

One more metric that need to be taken into consideration is the complexity c of a fragment/variant. We consider that complexity depends on the fragmentation ϕ that represents the number of pages hosting the fragment/variant. For a given fragment/variant, we compute its complexity as:

$$c_j = \frac{1}{\varphi_j} \qquad (5)$$

Overall, the impact (I) of a fragment/variant in the hypertext schema combines all the above metrics into a single metric according to the following computation:

$$I_j = f_j \times l_j \times c_j \qquad (6)$$

For each fragment/variant, in order to evaluate their impact on the hypertext schema, we also compute another metric:

$$R_j = \frac{I_v}{I_f} \times AverageSemSimScore \qquad (7)$$

I_f stands for the fragment's impact factor and I_v for the variant's impact factor. The higher the value of R_j, the better the appropriate of reuse is.

After estimating the value of the factor R_j for every occurrence of the identified fragments and variants, the results are presented in a descending order for both of the two aforementioned categories. Upon the completion of the above steps, developers can have access to a set of design fragments classified in two categories, one for the bad cases of reuse and one for the effective reusable design solutions.

4   Exemplifying Paradigm

Due to space limitations, we will exemplify the methodology by using a simple example referring to an instance of a socially network enabled website built using the Joomla! v. 2.5 platform, called TS-Eurotrain[1], in order to illustrate the discovery of a

[1]   Available at http://www.ts-eurotrain.eu/

reusable design solution in its hypertext schema. TS-Eurotrain is a website for publishing the research activity of a scientific network which studies the Tourette syn-
drome. The website publishes among others information concerning the projects undertaken through the network's participants and information about the members of the
teams. In the example depicted in figures 4 and 5, we capture a design solution which
occurs in two different places in the website's site view. The first (Figure 4) is a
fragment of the site view depicting a Projects area (supporting the categorization of
the implemented projects as industrial, research, etc.) while the second (Figure 5) is a
fragment depicting a member's area, supporting the categorization of the team's
members (network participants, external associated partners, etc.).

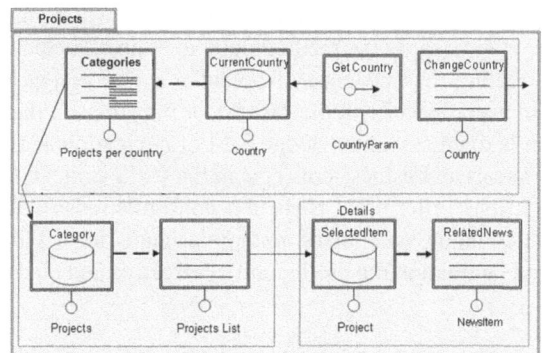

Fig. 4. A fragment of the Projects area site view

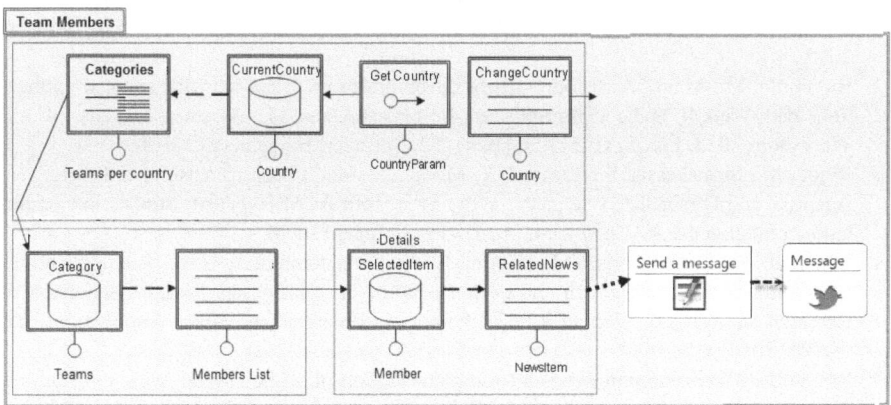

Fig. 5. A fragment of the team members area siteview

Comparing the two site view's fragments, we can easily identify the existence of an
effective reusable design solution. It consists of the entire composition of WebML
elements contained in the first fragment. When applying the methodology, this design
fragment is stored in the patterns repository. Moreover, in the case of Team Members
site view fragment, one variant of the identified design solution can be extracted, extending it with a Twitter message pattern. This way, we have possible identified a design solution for categorizing website's information items and their sharing on Twitter.

5   Conclusions and Future Work

In this paper, we have illustrated a model-driven approach for evaluating the design of Joomla! based web applications under the perspective of exploiting the social networking features. We provide an automatic way for extracting the hypertext schema of a web application which then is submitted to an analysis process in terms of the incorporated recurrent social design fragments implying design reuse. To evaluate the appropriateness of the reuse, we utilize a set of quality metrics categorizing the identified social design fragments as cases that either imply the possibility of refactoring the application or indicate the existence of effective reusable design structures that can lead to the discovery of new social design patterns in CMS domain.

In the future, we plan to apply the methodology to a remarkable number of Joomla! websites for investigating the existence of potential social design patters within the hypertext schema in accordance with the process of patterns and their variants identification. Our vision is to create a knowledge base of navigation and interface patterns that seem to perform certain business concepts in the CMS context. We plan to identify patterns specifications in terms of particular configurations of WebML hypertext elements forming a common vocabulary among designers for solving common CMS design problems and facilitating the production of effective and quality CMS designs.

References

1. Ceri S., Fraternali P., Bongio A.: Web modeling language (WebML): a modeling language for designing web sites. In: Proc. WWW Conference, Amsterdam, NL, pp. 137–157, May 2000
2. Brambilla, M., Mauri, A.: Model-driven development of social network enabled applications with WebML and social primitives. In: Grossniklaus, M., Wimmer, M. (eds.) ICWE Workshops 2012. LNCS, vol. 7703, pp. 41–55. Springer, Heidelberg (2012)
3. Rigou, M., Sirmakessis, S., Tzimas, G.: Model cloning: a push to reuse or a disaster? In: Adaptive and Personalized Semantic Web: Proc. 16th ACM Hypetext, Studies in Computational Intelligence (SCI), vol. 14, pp. 37–55. Springer (2006)
4. Fraternali, P., Tisi, M., Silva, M., Frattini, L.: Building community-based web applications with a model-driven approach and design pattern. In: Murugesan, S. (ed.) Handbook of Research on Web 2.0, 3.0, and X.0: Technologies, Business, and Social Applications. IGI Global (2010)
5. Brambilla, M.: From requirements to implementation of ad-hoc social Web applications: an empirical pattern-based approach. IET Software 6(2), pp. 114–126. doi:10.1049/iet-sen.2011.0041
6. Joomla! CMS platform. http://www.joomla.org/
7. Yan, X., Han, J.: gSpan: graph-based substructure pattern mining. In: ICDM 2002, Washington, DC, p. 721. IEEE Computer Society (2002)
8. Fraternali, P., Matera, M., Maurino, A.: WQA: an XSL framework for analyzing the quality of web applications. In: The Proceedings of the 2nd International Workshop on Web-Oriented Software Technologies–IWWOST 2002, Malaga, Spain, pp. 46–61 (2002)
9. ParSeMis – the Parallel and Sequential Mining Suite. https://www2.cs.fau.de/EN/research/zold/ParSeMiS/index.html

An e-Recruitment System Exploiting Candidates' Social Presence

Evanthia Faliagka[1(✉)], Maria Rigou[2,3], and Spiros Sirmakessis[1]

[1] Department of Computer and Informatics Engineering, Technological Institution
of Western Greece, National Road Antirrio-Ioannina 30020, Antirrio, Greece
faliagka@ceid.upatras.gr
[2] Department of Computer Engineering and Informatics, University of Patras,
Rion Campus, Patras 26500, Greece
[3] Hellenic Open University, Parodos Aristotelous 18 26335, Patras, Greece

Abstract. Applicant personality is a crucial criterion in many job positions. Choosing applicants whose personality traits are compatible with job positions has been shown to increase their satisfaction levels, as well as the rate of employee retention. However, the task of assessing candidates' personality is not addressed in today's online recruitment systems, but is typically handled during the interview process. The rapid deployment of social web services has made candidates' social activity much more transparent, giving us the opportunity to infer features of candidate personality with web mining techniques. In this work, a novel approach is proposed and evaluated for automatically extracting candidates' personality traits based on their social media use.

Keywords: e-recruitment systems · Personality mining · Social web mining

1 Introduction

The amount of increased information at all levels of people's electronic social environment has rapidly in the last years due to the expansion of the social web user base and frequency of recorded user activity in the so called social media [1]. Among the reasons that bring people to the web is knowledge and skills improvement [2], as well as career development [3]. Lately, people seeking for a job are increasingly using web 2.0 services like LinkedIn and job search sites [4], while a lot of companies also use web-based knowledge management systems to hire employees. These are termed e-recruitment systems and automate the process of publishing open job positions and receiving candidates' CVs. Online recruitment can be seeker-oriented or company-oriented. In the first case the e-recruitment system recommends to the candidate a list of job positions that better fit his profile. In the latter, recruiters publish the specifications of available job positions and the candidates can apply.

Typically in on-line recruitment systems, candidates upload their CVs in the form of a document with a loose structure, which must be considered by an expert recruiter. However this incorporates a great asymmetry of resources required from candidates and recruiters and potentially increases the number of unqualified applicants. This

© Springer International Publishing Switzerland 2015
F. Daniel and O. Diaz (Eds.): ICWE 2015 Workshops, LNCS 9396, pp. 153–162, 2015.
DOI: 10.1007/978-3-319-24800-4_13

situation might be overwhelming to HR agencies that need to allocate human re-sources for manually assessing the candidate resumes and evaluating their suitability for the positions at hand. Several e-recruitment systems have been pro-posed with an objective to automate and speed-up the recruitment process, leading to a better overall user experience and increasing efficiency. For example, after deploying an e-recruitment system SAT telecom reported 44% cost savings and a drop in the average time needed to fill a vacancy from 70 to 37 days [5].

In practice though, the required skills of the applicant are not the only factor that determines the final decision about hiring him or not, as personality is also considered crucial in many job positions. Candidates' personality is assessed by human recruiters at the interview stage, and is usually limited to candidates that passed the screening phase. However, it is known that the vast majority of recruiters perform a preliminary background check on applicants, based on their (social) web presence. This approach has limitations, as it is still a time consuming process and does not work well when the applicants use pseudonymous accounts or have a very common name and sur-name. A better approach would be to automate this background check, so that this background check could be performed by the e-recruitment system.

To this direction in [6] an integrated company oriented e-recruitment system is proposed. The system automates the candidate pre screening process providing an overall candidate ranking based on a combination of supervised learning and semantic skills matching. Applicant evaluation is based on a predefined set of objective criteria evaluated on the basis of applicant's skills that are directly extracted from his Linke-dIn profile, as well as his personality traits extracted by textually analyzing blog posts. One drawback of the aforementioned approach is that the popularity of the blog as medium of self-expression has been steadily declining. Instead, web users have turned to micro-blogging platforms that function with "status updates" instead of free text, such as Facebook, twitter, and Instagram.

In this work, we propose a different approach to infer a candidate's personality traits from his social web activity. More specifically, the personality characteristics that are assessed by the new system regard the candidate's interests as expressed by his social presence, how collaborative he is, whether his interests are related to his line of work and if he demonstrates leadership skills and is an influential individual for his electronic social environment.

The proposed company oriented e-recruitment system provides automated appli-cant ranking and personality mining with the purpose of restricting interviewing and background investigation of applicants to the top candidates identified by the system. This can lead to a major positive impact on the efficiency of the recruitment process and to significant cost savings. To showcase the effectiveness of the proposed scheme, a prototype has been implemented. The produced candidates' listing combin-ing character features automatically extracted from each candidate's social media activity has been assessed against actual personality traits manually evaluated by human recruiters. The next sections provide details regarding the architecture of the proposed system comprising three main modules, followed by the description of the new personality module. The discussion moves to the pilot recruitment scenario that was used to evaluate the effectiveness of the proposed approach and the paper ends with the main conclusions reached.

2   Architecture

Applicant personality is considered crucial in many job positions. It is assessed by human recruiters at the interview stage and is usually limited to candidates that passed the screening phase. However, it is a common practice among the majority of recruiters to perform a preliminary background check on applicants, based on their (social) web presence. This approach has limitations, as it is still a time consuming process and does not work well when the applicants use pseudonymous accounts or have a very common name and surname. A better approach would be to automate this background check, so that it is performed by the e-recruitment system. In [7] the authors proposed a first approach to such a system that performs automated extraction of candidate personality traits based on linguistic analysis of blog posts and automates the candidate evaluation and pre-screening process. One drawback of this approach is that it relies on the LIWC system to perform a textual analysis of candidates' blog posts to infer the "Big-5" personality directions. However the popularity of the blog as a medium of self-expression has been steadily declining. Instead, web users have turned to micro-blogging platforms that use "status updates" instead of free text, such as Facebook, Twitter and Instagram. Short "status" messages are typically accompanied with images and / or URLs to share information. Although a positive / negative tag can still be attributed to each status update [8] [9] [10], these lack the linguistic markers required by LIWC.

In this work, we propose a different architecture for extracting candidates' personality traits based on their social media use. An overview of the proposed system architecture is depicted in Fig. 1, and consists of the following components:

- *Job Application* Module: It implements the input forms that allow the candidates to apply for a job position. The candidate is given the option to log into the system with the account credentials of his micro-blogging platform of choice (i.e., either Twitter of Facebook).
- *Personality Mining* Module: If the candidate opts to log into the system with his Facebook or Twitter credentials, the system gains access to his "timeline" (i.e., stream of status updates) and interactions with other members of the community.
- *Applicant Grading* Module: The applicant's score in a set of predefined criteria is calculated. Web mining techniques are employed and applied to the candidate's social activity.

Although this is out of scope of this work, as can be seen in Fig. 1 a fully functional system is expected to also implement a ranking function that will assess the candidates' overall relevance to a specific job position, based on the scores of individual selection criteria. Numerous ranking functions based on AHP [11] or Machine Learning techniques [12] can be found in the literature.

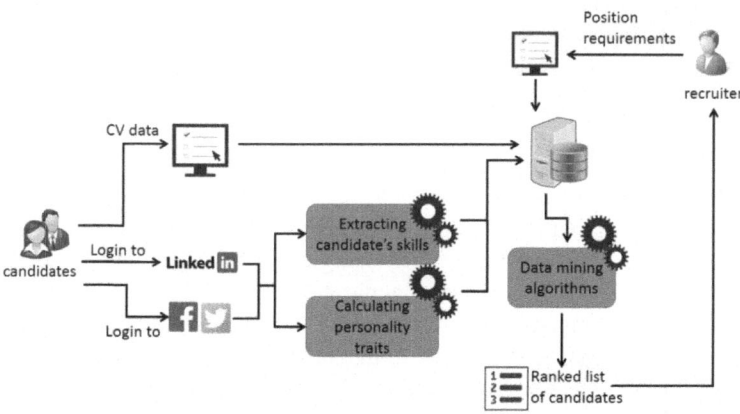

Fig. 1. System's architecture

3   The New Personality Module

An inherent limitation of automated recruitment systems is their over-reliance on formal qualifications. Hiring managers are able to see beyond skill-sets, and assess candidates' personality and how they would fit in the corporate culture. Judging applicants' creativity, inspiration or their ability to work with people is a hard problem to solve for automated systems (assuming that it can be solved at all). A new approach is proposed in this section. The proposed system is able to make inferences on the applicants' suitability for a specific position, by inferring his personality traits based on his social media use. Specifically, asking the applicant to log-in to the system with his Facebook or twitter credentials, we get access to a large amount of information, such as status updates, interaction with other users and interests ("likes") or side projects. These can demonstrate his strengths and passions in a way that the employment history can't always do [13]. For example, an individual involved in the organization of community events could boost an applicants' suitability for positions where organizational skills are critical. But even activities that do not directly relate to a specific position can translate to a well-rounded individual who takes time to "refuel and recharge" after work. In the following table (Table 1) we summarize the main personality traits that are of interest to the hiring managers, and how they are related to candidates' hobbies and leisure activities.

Table 1. Personality traits and the according activities

The candidate is a team player	Participation in a team sport or collaborative activity could translate to a person that is able to function in an activity that requires group interaction.
The candidate has leadership skills	Candidates that possess a leadership role in their social interactions are favored in management positions.
Actively working to improve his skills	Candidates that work on learning a new language or participate in seminars about public speaking are viewed positively.
The candidate is passionate	Good candidates are passionate about their activities, whether inside or outside of the office. Leisure activities relative to the position are always a plus
He is a well-rounded individual	Well-rounded personalities have an array of interests and are not merely focused on work.

3.1   Method

To assess personality we define the following criteria that quantify aspects of candidates' personality and can be mapped to real numbers in the interval [0, 1]. Their values can be calculated with web mining techniques, exploiting the candidates' social presence with respect to the following four criteria:

- *well-rounded*: quantifies in what degree a candidates' personality is fully developed, which is indicated by a number of non-work related leisure activities. Well-rounded candidates are persons with good social skills and are expected to function adequately in a team.
- *cooperative*: indicates well-rounded individuals with a demonstrated capacity to cooperate, as indicated by their participation in social after-work activities.
- *passionate*: candidates who are involved in activities related to the job position, or actively working in sharpening their skills after work.
- *influential*: candidates that others tend to use as a source of information or arguments. They tend to receive a high number of social interactions (e.g., re-shares, comments, etc.).

It is evident that these four criteria are not independent from each other. Rather, cooperative individuals are actually a subset of well-rounded personalities, while influential individuals are typically passionate persons that have the tendency to reveal information about their passions and promote them to other people.

The first step required to calculate the abovementioned selection criteria is to define a set of leisure activities that are supported by the system. These were selected from the categories of the Open Directory Project (ODP20) [14] and are shown in the following table (Table 2). Activities shown in bold are the collaborative activities that can indicate social and pleasant candidates.

Table 2. Activities that show a candidates' personality

1. Science	**2. Sports**
5. Travel	4. Computers
7. News	**6. Computer-Board Games**
9. Arts	8. Business

The second step is to create a corpus of words which are representative for each category. To create the corpus we used the Open Directory Project predetermined hierarchy of categories so as to access websites of a specific topic that respond to specific information needs. Then, we parsed these websites to make a corpus of words for each topic. Specifically, we used features extracted from the HTML code of the webpage including the pure text contained in it and the specific tag values as headings (<title>, <h1>) and meta information (<meta>). By discarding terms that appear less than 100 times (process only adjectives, verbs and nouns) we form a vocabulary that represents each category.

To find which categories are mentioned by the candidate often we analyze not only the hashtags that can directly give the information we need, but also the raw text of tweets and Facebook posts. We process the text in the tweets and Facebook posts and compute daily unigram frequencies. By discarding terms that appear less than 100 times, we form a vocabulary of size $|V| = 71, 555$. We then form a user term-frequency matrix with the mean term frequencies per user during the time interval Δt. All term frequencies are normalized with the total number of tweets and Facebook posts posted by the user. The final step is to compute a topic score for each user-topic pair to assign the interest categories to the candidates. For that reason, we used the Jaccard index, also known as the Jaccard similarity coefficient [15]. The Jaccard coefficient measures similarity between finite sample sets and is defined as the size of the intersection divided by the size of the union of the sample sets:

$$J(A, B) = \frac{|A \cap B|}{|A \cup B|} \quad (1)$$

Clearly,

$$0 \leq J(A, B) \leq 1 \quad (2)$$

To calculate the "well-rounded" metric we calculate the Jaccard coefficient for every pair of {activity words, candidate words} and the score is the greatest of the calculated values. In the same way we calculate the "cooperative score" and the "passionate score" using the corresponding activity words.

3.2   Identification of User Impact

As mentioned in previous sections, candidates that demonstrate passion for their leisure activities are also regarded more likely to be passionate in the workplace. One special category of passionate individuals are the ones that tend to influence others, attracting interest for their hobbies and leisure activities. In the social web, this is translated in a high degree of interactions with others. The most significant form of endorsement in the social media is the "share" or "retweet" function, in Facebook and Twitter platforms respectively. Influential users receive a high number of shares (i.e. other users repeat influential users' status updates on their own timelines). It must be noted that a high number of friends or followers is not a reliable metric of user impact, as shown in [16]. On the other hand, interactions (and especially shares) are indicative of a person that attracts interest and is viewed by others as a source of interesting information.

In order to quantify the user impact, we only take into account "status updates" (or "tweets") that fall into one of the categories illustrated in Table 2. The methodology detailed in Section 3.1 is employed to filter-out status updates or tweets that are not relevant to candidate interests. The rationale is that we are not interested in the candidate's impact as a whole in the social web, but rather on how successful he is in disseminating information about his leisure activities. This will generally imply that they are influential individuals. Thus, candidates' influence score is defined as the percentage of relevant status updates or tweets that their friends or followers interacted with (i.e., shared or retweeted). Notice that in the proposed method there is no dependency with the overall number of friends or followers, as would be the case if we counted the aggregate number of shares or retweets.

4   Pilot Scenario

In this section we evaluate the effectiveness of the proposed personality mining approach as detailed in Section 3, in a pilot recruitment scenario. The system's performance is assessed based on how effective it is in assigning consistent personality scores to the candidates, compared to the ones assigned by human recruiters. In our pilot scenario we compiled a corpus of 100 random twitter users, that reported to be working in the financial sector in their twitter bio. We subsequently derived their personality scores using the methodology detailed in Section 3. These scores were cross-compared with scores assigned by an expert recruiter, who had access to the same tweets.

The performance of the proposed system is evaluated based on how effective it is in discriminating the top candidates of each category, and providing a rank that is consistent with the one provided by the human recruiters. Three metrics were used for comparing rankings; the simplest one is the overlap size of the top-k list selected by the system and the human recruiter for each job position, where k=25 corresponds to 25% of overall applicants. The second metric is the correlation coefficient (Spearman's rho) of the top-k candidates per category. The third metric is the mean absolute difference (ranking error) of top-k candidate's ranks. The performance metrics for all three positions can be seen in Table 3.

Table 3. Performance evaluation metrics per job position

	Top-k	Correlation	Ranking error
Well-rounded	16 (64%)	0.66	4,8
Social	18 (72%)	0.68	4,1
Passionate	19 (76%)	0.71	3,9
Influential	22 (88%)	0.79	3,1

The calculation of the influential score was based on more objective criteria and thus this metric performed better than the others, with a correlation coefficient of 0.79. On the other hand the metrics that are based on the candidate's social presence had the lowest performance. Nevertheless, our method was able to output a top-25 list that overlapped at least 64% (for the well-rounded metric) and the correlation reached 0,71 (for the passionate metric).

Finally, in Fig. 2 we represent the candidates with circles positioned in a 2D plane based on the two most important personality scores (social and influential), while the circle radius is proportional to the candidate overall score, assigned by the recruiter. It is evident that most highly ranked candidates are clustered in the top right quadrant (i.e. with high social and influential scores), which attests that our tool assigned high ranks to candidates with the desired personality.

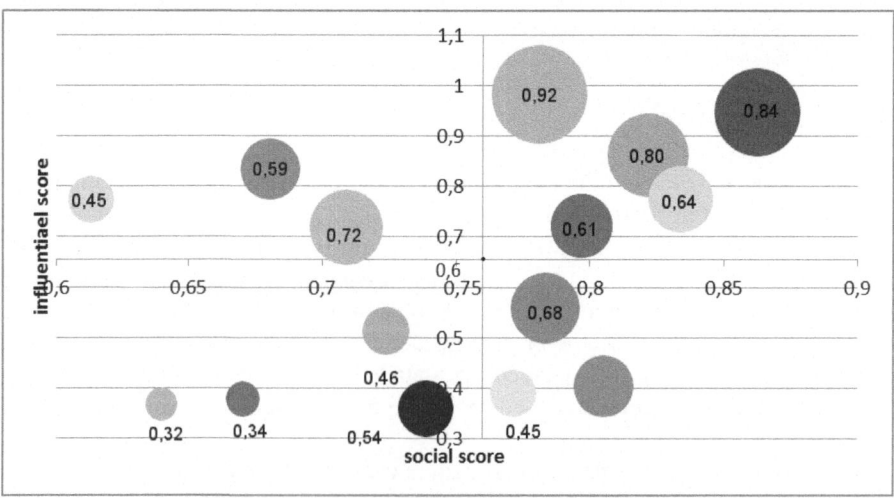

Fig. 2. The candidates' personality and overall scores

5   Conclusions

An inherent limitation of automated recruitment systems is their over-reliance on formal qualifications. Hiring managers are able to see beyond skill-sets, and assess candidates' personality and how they would fit in the corporate culture. Judging applicants' creativity, inspiration or their ability to work with people is a hard problem to solve for automated systems (assuming that it can be solved at all). This paper proposed a new approach in company-oriented recruitment by means of a system that provides automated candidates ranking and personality mining with the purpose of restricting interviewing and background investigation of applicants to the top candidates identified by the system. The proposed system is able to assess the applicants' suitability for a specific position, by inferring his personality traits based on his social media use. More specifically each candidate is evaluated based on how well-rounded, social, passionate and influential he is according to information collected by his social media activity. The performance of the system has been evaluated on the basis of how effective it is in discriminating the top candidates of each category, and outputting a ranking that is consistent with the one provided by the human recruiters. Among the four personality scores, the influential score obtained the highest accuracy, while the more subjective social score had the lowest. Regardless though of the partial scoring, it is quite positive that the system managed to output a top-25 list that overlapped with the corresponding ranking of the human recruiters at least 64% (for the well-rounded metric) and the correlation reached 0,71 (for the passionate metric). We plan to experiment with the system in full scale in order to investigate potentially better fine-tuning in the algorithm that assesses the four personality scores and set up recruitment scenarios in various domains.

References

1. Neuman, C.: Prospero: a tool for organizing Internet resources. Internet Research **20**, 408–419 (2010)
2. Ho, L., Kuo, T., Lin, B.: Influence of online learning skills in cyberspace. Internet Research **20**, 55–71 (2010)
3. Jansen, B., Jansen, K., Spink, A.: Using the web to look for work: Implications for online job seeking and recruiting. Internet Research **15**, 49–66 (2005)
4. Bizer, R.H., Rainer, E.: Impact of Semantic web on the job recruitment Process. Wirtschaftsinformatik, Physica-Verlag HD (2005)
5. Pande, S.: E-recruitment creates order out of chaos at SAT Telecom: System cuts costs and improves efficiency. Human Resource Management International Digest **19**, 21–23 (2011)
6. Faliagka, E., Iliadis, L., Karydis, I., Rigou, M., Sioutas, S., Tsakalidis, A., Tzimas, G.: On-line Consistent Ranking on e-Recruitment: Seeking the Truth Behind a Well-formed CV. Artif. Intell. Rev. **42**, 515–528 (2014)
7. Faliagka, E., Tsakalidis, A., Tzimas, G.: An integrated e-recruitment system for automated personality mining and applicant ranking. Internet Research **22**(5), 551–568 (2012)
8. Agarwal, A., Xie, B., Vovsha, I., Rambow, O., Passonneau, R.: Sentiment analysis of twitter data. In: Proceedings of the Workshop on Languages in Social Media. Association for Computational Linguistics (2011)

9. Kouloumpis, E., Wilson, T., Moore, J.: Twitter sentiment analysis: The good the bad and the omg! ICWSM **11**, 538–541 (2011)
10. Pak, A., Paroubek, P.: Twitter as a corpus for sentiment analysis and opinion mining. In: LREC, vol. 10 (2010)
11. Saaty, T.L.: How to make a decision: The analytic hierarchy process. European Journal of Operational Research (1990). Elsevier Science B.V.
12. Basak, D., Srimanta, P., Dipak, C.P.: Support vector regression. Neural Information Processing-Letters and Reviews **10**, 203–224 (2007)
13. http://www.businessinsider.com/why-job-interviewers-ask-about-hobbies-2014-8#ixzz3 VCFcHCja
14. http://www.dmoz.org/
15. Suphakit, N., Jatsada, S., Ekkachai, N., Supachanun, W.: Using of jaccard coefficient for keywords similarity. In: Proceedings of the International MultiConference of Engineers and Computer Scientists, vol. 1 (2013)
16. Cha, M., Haddadi, H., Benevenuto, F., Gummadi, P.K.: Measuring User Influence in Twitter: The Million Follower Fallacy. ICWSM **10**, 10–17 (2010)

#Nowplaying on #Spotify: Leveraging Spotify Information on Twitter for Artist Recommendations

Martin Pichl$^{(\boxtimes)}$, Eva Zangerle, and Günther Specht

Databases and Information Systems, Institute of Computer Science,
University of Innsbruck, Innsbruck, Austria
{martin.pichl,eva.zangerle,guenther.specht}@uibk.ac.at

Abstract. The rise of the web enabled new distribution channels like online stores and streaming platforms, offering a vast amount of different products. For helping customers finding products according to their taste on those platforms, recommender systems play an important role. Besides focusing on the computation of the recommendations itself, in literature the problem of a lack of data appropriate for research is discussed. In order to overcome this problem, we present a music recommendation system exploiting a dataset containing listening histories of users, who posted what they are listening to at the moment on the microblogging platform Twitter. As this dataset is updated daily, we propose a genetic algorithm, which allows the recommender system to adopt its input parameters to the extended dataset. In the evaluation part of this work, we benchmark the presented recommender system against two baseline approaches. We show that the performance of our proposed recommender is promising and clearly outperforms the baseline.

Keywords: Music recommender systems · Collaborative filtering · Social media · Twitter

1 Introduction

The way how consumers access music has changed in recent years due to the rise of the web. Nowadays, consumers have the possibility to access a huge amount of different music using various devices and services. An example for such new services are music streaming platforms. The distribution and especially the inventory costs of such platforms are lower than the costs of traditional channels, i.e., brick and mortar stores. Due to this development, an increased amount of more diverse music is available, as additionally to bestsellers also niche music can be offered with low additional costs [1]. Besides commercial vendors like Spotify[1] or Pandora[2], there are also open platforms like SoundCloud[3] or Promo DJ[4],

---

[1] http://www.spotify.com
[2] http://www.pandora.com
[3] http://soundcloud.com
[4] http://promodj.com

© Springer International Publishing Switzerland 2015
F. Daniel and O. Diaz (Eds.): ICWE 2015 Workshops, LNCS 9396, pp. 163–174, 2015.
DOI: 10.1007/978-3-319-24800-4_14

where users can upload and publish their own creations, increasing the diversity
of available music. In contradiction to traditional channels like the radio, these
new channels allow consumers to freely choose tracks they listen to and to create
own playlists. One drawback of this freedom of choice is, that it is difficult for
the customers to identify new tracks they like and want to listen to in this sheer
amount of artists and tracks. As explained in Section 2, recommender systems
can be seen as method to deal with information overload and a type of searching for information [7]. Thus, implementing a recommender system helps users
discovering new tracks according to their taste, which increases user satisfaction.

As most data corpora in this field are owned by private companies, e.g.,
Spotify or Pandora mentioned above, this data is not or only publicly available limited. Thus, we propose to utilize Twitter and Spotify as new sources
of publicly available data, which enables us to develop and evaluate our recommender system. Furthermore, we steadily enlarge the dataset. Since July 2011,
the Research Group Databases and Information System of the University of
Innsbruck is crawling tweets via the Twitter Streaming API. The whole process
generating a dataset suitable for music recommendations and the basic statistics
are presented in Section 3.1. In order to foster research in this field, the dataset
is made available publicly[5].

Our main contribution, beside the dataset itself, is that it is possible to provide music recommendations by utilizing Twitter data. We propose an approach
which adopts itself to this ever changing data. We show that a promising solution for dealing with these changes is a genetic logarithm (GA). Beside this, the
evaluation of this recommender system shows, that in the case of CF, the track
listening history seems to be more important than the artist listening history of
a user.

In short, this paper addresses two problems: The first problem is the recommendation process itself. This problem is addressed by a collaborative filtering
(CF) based recommender system, which is presented in more detail in Section 2.
Afterwards, in Section 3, the second problem is addressed, which is a lack of
publicly available and recent data appropriate for implementing and evaluating
recommender system in academia [12]. This problem is addressed by exploiting
Twitter and Spotify as a source of publicly available data. We briefly describe
the generation of the dataset and the dataset itself before focusing on the evaluation in Section 3. Before we present the conclusions drawn from the evaluation
in Section 5, related work is discussed in Section 4.

2   The Recommender System

In this section, the implementation of the artist recommender system is described
in more detail. However, firstly we briefly explain the purpose of recommender
systems and introduce the reader to the methods used.

---

[5] available at: http://dbis-twitterdata.uibk.ac.at/spotifyData/

2.1   Background

In general, recommender systems can be seen as a method to deal with information overload [7] and a new type of searching for information. A common method for overcoming the information overload is to state one recommendation or a list of recommendations, which is much shorter than the initial list of all potential candidate items [7]. This is exactly what we do: Our recommender system helps users finding music they want to listen to or buy, by stating a short list of recommendations. In particular, our system provides users with artist recommendations, a user might be interested in.

In order to provide these artist recommendations, we rely on user-based collaborative filtering (CF). User-based CF recommends items by finding similar users, herein referred to as the nearest neighbours, and then suggesting items a user didn't interacted with, but his nearest neighbours interacted with. This is based on the assumption that like-minded users prefer the same items and that these user preferences remain stable over time [12]. For the music recommender system presented in this work, a user-item interaction states that a user listened to a certain track by a certain artist. Thus, we recommend artists the nearest neighbours of a user listened to but are new to the user himself.

A recommender system as ours, relies on certain input parameters, i.e., the number of nearest neighbours taken into consideration for retrieving the recommendations. In order to estimate suitable parameters, we implemented a genetic algorithm (GA) [10] as a possible solution. A GA is a heuristic for solving search and optimization problems [14]. The decision for performing the parameter optimization using GA was taken, as it would be computationally too expensive to compute the precision of our recommender system with all possible parameter combinations. The suitability of a GA for overcoming the problem of selecting and weighting input parameter, amongst others for recommender systems, was shown by Fong et al. [6]. The idea behind a GA, as introduced by Holland [10], is that the chromosomes of the fittest individuals of a population are inherited to the next generation. This inheritance, also called natural selection is achieved by recombining the genes of the fittest to new chromosomes by crossovers. Additionally, they are altered by random mutations and inversions [14].

How our artist recommender system computes the recommendations, on which input parameters it relies and how they are optimized is presented in the consecutive section.

2.2   Implementation

An overview of the recommendation process is shown in Figure 1. Using the Twitter Streaming API, a crawler is continually crawling for new tweets containing one of the following keywords: *nowplaying, listento* and *listeningto*. The update of the dataset, which means that the new tweets are resolved using the contained Spotify URL and stored in a database, can be done periodically. In our case this is done every 24 hours. This implies that our dataset is steadily enlarged and the optimal parameters may change. Thus, these parameters have to be updated along with the dataset. Therefore, an update of the dataset triggers the genetic algorithm (GA) which is updating the parameters for the music recommender system. This is why the actual recommendation process consists of two steps: In the first step, good input parameters have to be found, before providing actual recommendations using collaborative filtering (CF) in the second step.

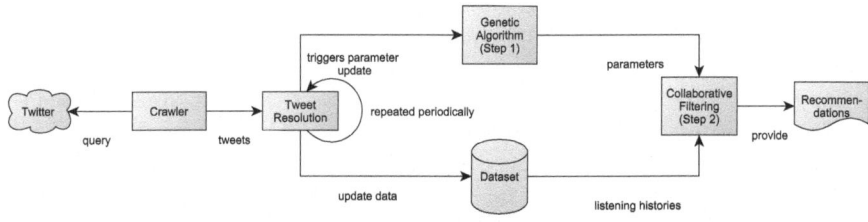

Fig. 1. The Recommendation Workflow

In the following, we describe our CF approach and present the input parameters the recommender system depends on. As already mentioned, User-based CF recommends items by finding similar users: the nearest neighbours. In our scenario, user listening histories are used for identifying the nearest neighbours and then computing the actual artist recommendations. As the artist listening history we consider the set of all artists a user i listened to ($artists_i$) in the training data and analogously, we consider the set of all tracks a user listened to as the track listening history ($tracks_i$). In order to compute the user similarities for determining the nearest neighbours, we applied the Jaccard Coefficient, as defined in Equation 1 [11], to the listing histories of two users i and j.

$$jaccard_{i,j} = \frac{|S_i \cap S_j|}{|S_i \cup S_j|} \tag{1}$$

The Jaccard Coefficient measures the similarity of two sets S between 0 and 1. We compute this metric for the artist- as well as a track similarity as defined in Equations 2 and 3:

$$artistSim_{i,j} = \frac{|artists_i \cap artists_j|}{|artist_i \cup artists_j|} \tag{2}$$

$$trackSim_{i,j} = \frac{|tracks_i \cap tracks_j|}{|tracks_i \cup tracks_j|} \tag{3}$$

The final user similarity ($userSim$) is computed as the weighted average of both Jaccard Coefficients as depicted in Equation 4.

$$userSim = w_a * artistSim + w_t * trackSim. \tag{4}$$

The weights w_a and w_t determine the influence of the artist- and the track listening history on the user similarity, where $w_a + w_t = 1$. Thus, if $w_t = 0$, only the artist listening history is taken into consideration and analogously, if $w_a = 0$, only the track listening history is taken into consideration.

Using this user similarity, we are able to compute the similarity of a user i to all other users in the dataset and determine the k-nearest neighbours. These k-nearest neighbours are used to derive a set of possible recommendations. These recommendation candidates are items the nearest neighbours listened to, but have not been listened to by the user i.

For ranking the candidate items, we implemented two strategies: The first strategy (rs_1) computes a ranking coefficient r for every item i by counting how often an item occurs among the k-nearest neighbours, whereas the second strategy (rs_2) computes r by summing up the similarities of user i to any other user j found among the k-nearest neighbours K, who interacted with the same item as shown in Equation 5.

$$r_i = \sum_{j \in K} userSim_{i,j} \qquad (5)$$

In order to perform the computation presented above, which is the second recommendation step, the following four parameters have to be set in the first step:

1. The weight of the *artistSim* : w_a
2. The weight of the *trackSim* : w_s
3. The number of nearest neighbours k
4. The ranking strategy R

These parameter form the genes of the GA. In order to build a initial population and allow mutations we generate random values for the float point and integer genes (w_a, w_t, k) and bit flips for the boolean gene (R). The *fitness* of an individual (in this case a recommender system) in the population is defined by the *precision* of this individual. We use the *precision* as the fitness measure, as we want to compute a short list of recommendations and to find all artists a user likes. For this so-called "find good items task", this is a reasonable metric [9]. Each individual has four genes forming the chromosome. The chromosome encodes the presented recommender parameters w_a, w_t, k and R. The *precision* of each individual is computed as described in Section 3.2.

These parameters should remain stable, as long as there are no changes in the dataset. Thus, the costly estimation of the parameters does not have to be performed prior each recommendation, but every time there is a change to the dataset, i.e., if it is updated as shown in Figure 1. After presenting the implementation of our recommender system, we present how the evaluation of this recommender system is conducted in the next section.

3   Recommendation Evaluation

In this section, the used dataset as well as the details of the recommender system evaluation are presented. As already mentioned, the evaluation is used to find suitable input parameters for the recommender system but also to show the performance of the recommender system.

3.1   Dataset

In contrast to other works aiming at extracting music information from Twitter, where the tweet's content is used to extract artist and track information from [8,15,19], we propose to exploit the subset of crawled tweets containing a URL leading to the website of the Spotify music streaming service. I.e., information about the artist and the track are extracted from the website mentioned in the tweet, rather than from the content of the tweet. This enables an unambiguous resolution of the tweets, in contradiction to the works mentioned above, where the text of the tweets is compared to entries in the reference database using some similarity measure. A typical tweet, published via Spotify, is depicted in the following: "#nowPlaying I Tried by Total on #Spotify http://t.co/ZaFH ZAokbV", where a user published that he or she listened to the track "I Tried" by the band "Total" on Spotify. Additionally, a shortened URL is provided. Besides this shortened URL, Twitter also provides the according resolved URL via its API. This allows for directly identifying all Spotify-URLs by searching for all URLs containing the string "spotify.com" or "spoti.fi". By following the identified URLs, the artist and the track can be extracted from the title tag of the according website.

Besides an unambiguous resolution and thus as clean dataset, utilizing Spotify tweets enables us to crawl additional information from Spotify about the track and about the user for future works. More details about this is presented in the outlook in Section 5.

The evaluation dataset is based on a snapshot of the dataset taken on April 30th, 2015 and is available online. This snapshot contains [513,489] unique $< user, \ artist, \ song >$-triples. Among the dataset we found 97,586 unique tracks by 40,593 artists, listened by 68,045 different users. One issue, which is addressed in a future work by user tweets crawling, is the data sparsity of the dataset. The sparsity can be seen in in Table 1, where the number of user with a minimum amount of tweets in the dataset is stated. Data sparsity is an issue, as normally the performance of CF based recommender systems increases with the detailedness of a user profile. This problem of the data sparsity will be addressed in a future work, as described in Section 5.

3.2   Evaluation Setup

In order to assess the performance of the recommender system, we conducted an offline evaluation. The details of this evaluation are presented in this section.

Table 1. Number of Tweets and Number of Users

Number of Tweets	Number of Users
> 0	68,045
> 1	38,906
> 10	9,111
> 100	516
> 1,000	12

For computing the *precision* as depicted in Equation 6, we split the listening history of each user into a training- and a test set. The test set was created by randomly removing one third of the listening events. The other two thirds of the listening events were actually used to compute the recommendations. If a computed recommendation was found in the test set, it was considered as a true positive.

$$precison = \frac{True\ Positives}{Number\ of\ Recommendations} \tag{6}$$

Furthermore we computed the *recall* as shown in Equation 7. For computing the *recall* metric, all items in the test set are considered as relevant items (and hence are desirable to recommend to the user). The recall metric describes the fraction of relevant artists who are recommended., i.e., when recommending 5 items, even if all items are considered relevant, the maximum recall is still only 50% if the size of the test set is 10. Thus, in this evaluation setup, *recall* is bound by an upper limit, which is the number of recommended items divided by the size of the test set.

$$recall = \frac{True\ Positives}{Size\ of\ the\ Testset} \tag{7}$$

Finally, for each user, we computed the *precision* and *recall* metrics for recommending a different number of artists depending on the size of the test set as depicted in Equation 8. We varied the percentage value p between 10% up to 100% and repeated each experiment 10 times in order to estimate the variance.

$$recommendations = p * Size\ of\ the\ Testset \tag{8}$$

3.3 Detection of Optimal Parameters

As stated in Section 2, for finding good input parameters a GA [10] is implemented. To measure the fitness of a population, we used the *precision* as argued in Section 2.2. The size of the population was set to 10 individuals. As depicted in Figure 2, with this configuration and using a snapshot of the dataset taken on April 30th, 2015, a good solution is found after 6 iterations. On average, a solution was found after 4.14 iterations ($SD = 2.27$), thus we propose to terminate the recommender after 7 iterations ($M + SD = 4.14 + 2.27 = 6.41$).

This solution, a recommender system with the parameters as stated below, is used for the performance evaluation of the recommender system in the following section.

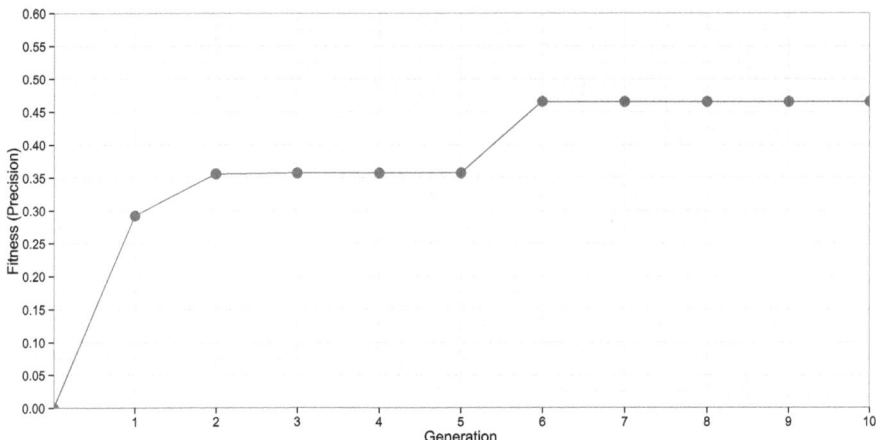

Fig. 2. Fitness (Precision) of the best Individual in each Generation

1. The weight of the $artistSim : w_a = 0.21$
2. The weight of the $trackSim : w_s = 0.94$
3. The number of nearest neighbours $k = 59$
4. The ranking strategy $R = rs_2$

From the result we can see that for the current snapshot of the dataset, the track listening history seems to be more significant for calculating the user similarity. Furthermore, we can also see that there was a mutation as $w_a + w_t \neq 1$.

3.4   Performance of the Recommender System

The parameters detected in Section 3.3 were used in order to perform the performance analysis of our proposed recommender system (new Approach). We compare the performance of our recommender system to two baseline recommendation approaches: Baseline approach number 1 (BA1) simply recommends the most popular items in the dataset to each of the users, whereas baseline approach number 2 (BA2) is a standard user-based CF approach without any further optimization than the number of nearest neighbours. Thus, the parameters of the second baseline recommendation approach are: $w_a = 1$, $w_t = 0$, $k = 59$. The performance of all three recommender systems is shown in Figure 3 and Table 2 (standard deviations in parentheses).

Although our proposed recommender system clearly outperforms both baseline approaches and thus is a promising approach, the *precision* drops rather fast with the number of recommendations (see Figure 3). This is, as the chance of false positives increases with the number of recommendations, if the size of the test set is kept constant. However, as discussed in Section 5, we already experimenting with further performance improvements by incorporating the user's context into the recommendation process.

Table 2. Precision and Recall of the Recommender Systems

p	BA1 Precision	BA1 Recall	BA2 Precision	BA2 Recall	new Approach Precision	new Approach Recall
0.1	0.156 (0.039)	0.024 (0.003)	0.316 (0.024)	0.015 (0.001)	0.466 (0.023)	0.022 (0.001)
0.5	0.117 (0.002)	0.026 (0.001)	0.117 (0.006)	0.060 (0.003)	0.253 (0.006)	0.120 (0.003)
1.0	0.081 (0.007)	0.032 (0.002)	0.096 (0.003)	0.081 (0.003)	0.181 (0.005)	0.158 (0.005)

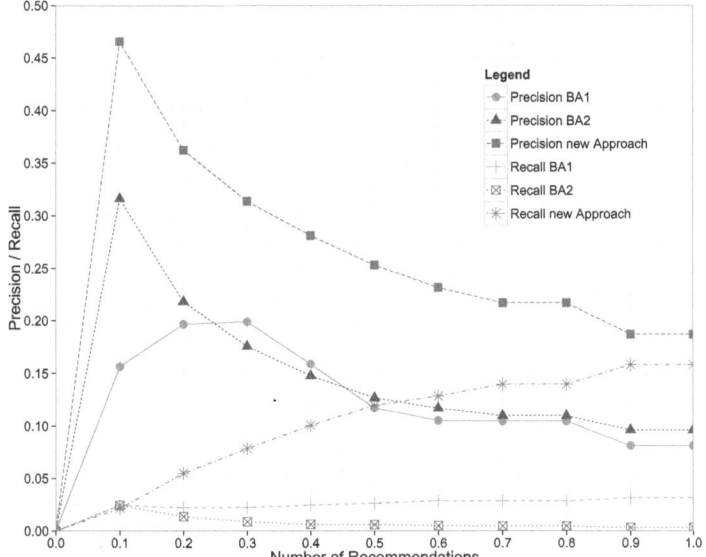

Fig. 3. Precision and Recall of the Recommender Systems

4   Related Work

Work related to the approach presented in this paper is concerned with both problem domains addressed in this work: (i) the generation of appropriate datasets for research in the field of recommender systems and (ii) research in the field of (music) recommender systems itself.

Regarding (i), Hauger et al. [8] published the Million Musical Tweets Dataset (MMTD). The MMTD is a static dataset containing about one million music related tweets. All tweets in this dataset are tagged with a geolocation. Similar to ours, their dataset also suffers from a high data sparsity or rather a low number of tweets per user. Another very popular dataset, containing one million listening events, is based on data logged by the music discovery service Last.fm. The Million Song Dataset (MSD) [2], released 2011, contains information about one million songs from different sources. Beside real user play counts, it provides audio features of the songs and is therefore suitable for CF, content-based (CB) and hybrid recommender systems. Besides those comparable large datasets, Yahoo! published numerous datasets[6] containing ratings for artists and songs suitable for CF. The largest Yahoo! dataset contains 136,000 songs along with ratings given by 1.8 million users. The data itself was gathered by monitoring users using the Yahoo! Music Services between 2002 and 2006 and thus, the dataset is less recent. Additionally to the ratings, the Yahoo dataset contains genre information which can be exploited by a hybrid recommender system. Celma, based on his research [4], also provides a music dataset, containing data retrieved from last.fm. It is comparably small and contains user, artists and play counts as well as the MusicBrainz identifiers for 360,000 users. This dataset was published in 2010 [4]. Another interesting dataset, which is in contradiction to ours suitable for recommending movies, is the MovieTweetings dataset[7] published by Dooms et al. [5]. As the dataset used in this work, it is continually updated and contains movies and the corresponding user ratings extracted from structured tweets tweeted via the Internet Movie Database (IMDb)[8].

Beside works concerned with generating datasets, related work is also concerned with recommendation approaches in general. There are various approaches for implementing recommender systems. Burke [3] classifies recommender systems by the data or rather the data source they use: CB recommendation approaches use information about items, whereas recommender systems based on CF take information about all users of a system or rather the user's taste into account. Demographic-based recommender systems use demographic data and Knowledge-based recommender systems rely on user needs. Finally, there are Hybrid recommender systems, combining the different approaches mentioned above. In the field of music recommendation, different hybrid approaches have been evaluated. Yoshii et al. state, that hybrid recommender system exploiting CF and CB data can outperform recommender systems based solely on CF [18] and Schedl et al. [16,17] argue, that a CF based recommender systems incorporating the geolocation works best.

Also GAs have been used for recommender systems in similar settings than ours. Hyun-Tae Kim et al. [13] propose a CB recommender system based on different audio features. They use an interactive genetic algorithm (IGA), where the users are evaluating the fitness of each population manually, in order to enable the system the music preferences of the users and recommend suitable music tracks. Instead of relying on a dataset, they conduct a real user experiment. They claim that their approach is a success, however the conducted experiment incorporates only ten users. Similar to us, Fong et al. [6] address the problem of finding input parameters using a GA. The use a GA for selecting and weighting different features of their CB recommender system suitable for recommending movies.

---

[6] available at: http://webscope.sandbox.yahoo.com/catalog.php?datatype=r
[7] http://github.com/sidooms/MovieTweetings
[8] http://www.imdb.com

5   Discussion and Future Work

The recommender system implemented in this work delivers promising results compared to the baseline approaches. From evaluations results stated in Table 2, we can derive that for the current snapshot of the dataset, the track listening history seems to be more significant for calculating the user similarity. We lead this back to the fact, that the track implicitly contains the artist. Besides this we see that at the current stage, the performance of the proposed recommender system is limited if the number of recommendations is high. In order to further boost the performance, we will aim at (i) enlarging the dataset by user playlist crawling on Spotify to reduce the data sparsity and (ii) incorporating the user's listening context. The idea behind (ii) is, that people are listening to different music in different situations. First experiments have shown, that incorporating the context is a promising approach: We already crawling additional information like favourite tracks and playlist names. From those playlist names, we are able to extract context-based information: We observed for instance playlists for certain events like Christmas, weddings or workouts but also playlist of certain moods, i.e., chill, relax, in love or the like. At the moment we are experimenting how to integrate this information into the presented recommender system.

6   Conclusion

Our proposed recommender system uses a genetic algorithm for optimizing the input parameters to the presented, steadily updated dataset. For computing the actual recommendations, it uses user-based collaborative filtering with a hybrid user-similarity computation. Although the implemented algorithm is simple, the performance of the recommender systems is promising and clearly outperforms the baseline approaches, namely recommending the most popular artists and standard user-based collaborative filtering. Nevertheless, as discussed in the previous section, there is space left for further performance improvements. Thus, we integrate context-based information in the next step in order to design a more specialized and more user-centric algorithm.

References

1. Anderson, C.: The Long Tail: Why the Future of Business Is Selling Less of More. Hyperion (2006)
2. Bertin-Mahieux, T., Ellis, D.P.W., Whitman, B., Lamere, P.: The million song dataset. In: Proceedings of the 12th International Society for Music Information Retrieval Conference (ISMIR 2011), pp. 591–596. University of Miami (2011)
3. Burke, R.: Hybrid recommender systems: Survey and experiments. User Modeling and User-Adapted Interaction **12**(4), 331–370 (2002)
4. Celma, Ò.: Music Recommendation and Discovery - The Long Tail, Long Fail, and Long Play in the Digital Music Space. Springer (2010)

5. Dooms, S., De Pessemier, T., Martens, L.: Movietweetings: a movie rating dataset collected from twitter. In: 7th ACM Conference on Recommender Systems Workshop on Crowdsourcing and Human Computation for Recommender Systems (RecSys 2013) (2013)
6. Fong, S., Ho, Y., Hang, Y.: Using genetic algorithm for hybrid modes of collaborative filtering in online recommenders. In: Proceedings of the Eighth International Conference on Hybrid Intelligent Systems (HIS 2008), pp. 174–179 (2008)
7. Goldberg, D., Nichols, D., Oki, B.M., Terry, D.: Using collaborative filtering to weave an information tapestry. Commun. ACM **35**(12), 61–70 (1992)
8. Hauger, D., Schedl, M., Kosir, A., Tkalcic, M.: The million musical tweet dataset - what we can learn from microblogs. In: Proceedings of the 14th International Society for Music Information Retrieval Conference (ISMIR 2013), pp. 189–194 (2013)
9. Herlocker, J.L., Konstan, J.A., Terveen, L.G., Riedl, J.T.: Evaluating collaborative filtering recommender systems. ACM Transactions on Information Systems **22**(1), 5–53 (2004)
10. Holland, J.: Adaptation in Natural and Artificial Systems. University of Michigan Press, Ann Arbor (1975)
11. Jaccard, P.: The distribution of the flora in the alpine zone. New Phytologist **11**(2), 37–50 (1912)
12. Jannach, D., Zanker, M., Felfernig, A., Friedrich, G.: Recommender Systems: An Introduction. Cambridge University Press (2010)
13. Kim, H.-T., Kim, E., Lee, J.-H., Ahn, C.W.: A recommender system based on genetic algorithm for music data. In: Proceedings of the 2nd International Conference on Computer Engineering and Technology (ICCET 2010), vol. 6, pp. V6-414–V6-417 (2010)
14. Mitchell, M.: An Introduction to Genetic Algorithms. MIT Press, Cambridge (1998)
15. Schedl, M., Schnitzer, D.: Hybrid retrieval approaches to geospatial music recommendation. In: Proceedings of the 35th Annual International ACM SIGIR Conference on Research and Development in Information Retrieval (SIGIR 2013) (2013)
16. Schedl, M., Schnitzer, D.: Location-aware music artist recommendation. In: Gurrin, C., Hopfgartner, F., Hurst, W., Johansen, H., Lee, H., O'Connor, N. (eds.) MMM 2014, Part II. LNCS, vol. 8326, pp. 205–213. Springer, Heidelberg (2014)
17. Schedl, M., Vall, A., Farrahi, K.: User geospatial context for music recommendation in microblogs. In: Proceedings of the 37th Annual International ACM SIGIR Conference on Research and Development in Information Retrieval (SIGIR 2014) (2014)
18. Yoshii, K., Goto, M., Komatani, K., Ogata, T., Okuno, H.G.: Hybrid collaborative and content-based music recommendation using probabilistic model with latent user preferences. In: Proceedings of the 7th International Conference on Music Information Retrieval (ISMIR 2006), pp. 296–301 (2006)
19. Zangerle, E., Gassler, W., Specht, G.: Exploiting twitter's collective knowledge for music recommendations. In: Proceedings of the 2nd Workshop on Making Sense of Microposts (#MSM 2012), pp. 14–17 (2012)

Retrieving Relevant and Interesting Tweets During Live Television Broadcasts

Rianne Kaptein[1]([✉]), Yi Zhu[2], Gijs Koot[1], Judith Redi[2], and Omar Niamut[1]

[1] TNO, The Hague, The Netherlands
rianne.kaptein@tno.nl
[2] Delft University of Technology, Delft, The Netherlands

Abstract. The use of social TV applications to enhance the experience of live event broadcasts has become an increasingly common practice. An event profile, defined as a set of keywords relevant to an event, can help to track messages related to these events on social networks. We propose an event profiler that retrieves relevant and interesting tweets in a continuous stream of event-related tweets as they are posted. In our application, these tweets are to be displayed in real time along with the live broadcast of an event. To test our application we have executed a user study. Feedback is collected during a live broadcast by giving the participant the option to like or dislike a tweet, and by judging a selection of tweets on relevancy and interestingness in a post-experiment questionnaire. From the experimental results we conclude that our event profiler is capable to find relevant keywords which leads to retrieval of a higher number of relevant and, on average, more interesting tweets than using only a manually selected keyword.

1 Introduction

Watching TV and using social media are not mutually exclusive activities. Many people are active on social networks while watching TV to discuss what they are watching, especially on Twitter [3,15]. To stay up to date with the event discussion in Twitter, while enjoying watching the event itself, people can use applications such as HootSuite[1] and Beamly[2]. These applications work by tracking the offical Twitter account and hashtag of the event; if the user wants to access more and diverse tweets, he or she will have to manually add a list of keywords to be tracked, including hashtags and usernames of interest.

Moreover, not all tweets containing the official hashtag are interesting to show to users. A tweet that simply states "watching icehockey #sochi 2014" may be of limited interest for the majority users who do not know the author of such tweet. Therefore, tracking the sole official hashtag may be suboptimal to retrieve interesting and relevant tweets for an event. In this paper we focus on live events, but a similar approach could be followed to track social media

---

[1] https://hootsuite.com/
[2] http://beamly.com/

© Springer International Publishing Switzerland 2015
F. Daniel and O. Diaz (Eds.): ICWE 2015 Workshops, LNCS 9396, pp. 175–185, 2015.
DOI: 10.1007/978-3-319-24800-4_15

activity around televisions shows, although the fans in this case might use official hashtags more often for a sense of community [15]. In this paper we investigate the following research question:

- How can we design an event profiler that generates a set of query terms to retrieve relevant and interesting tweets during an event?

We propose a novel event profiler which allows to retrieve relevant and interesting tweets in a continuous stream of event-related tweets as they are posted. We define an *event profile* as a set of keywords relevant to the event, which is used to track messages related to the event on social networks. This set of keywords includes not only hashtags but also frequent words and usernames, and contains the most relevant ones based on the similarity of their language model to that of a manually selected keyword (e.g. the official hashtag).

We present the results of an empirical investigation of the performance of such tools, which will answer the following two questions:

1. How does our event profiler perform in generating relevant and interesting tweets, compared to a traditional tool based on the simple event hashtag?
2. What is the relation between relevancy and interestingness of tweets? Information retrieval papers usually only consider relevancy as a performance measure, but for our application we also look at interestingness: are the users interested in seeing this tweet, do they like it?

The remainder of this paper is organized as follows. In the next section we discuss related work. In Section 3, we describe the design of our event profiler and its functioning within a social TV application that displays relevant tweets along with a live event broadcast. The user tests of the application are reported in Section 4. Finally, in Section 5 we draw our conclusions.

2   Related Work

Event profiling can be seen as a form of event tracking. A substantial part of the related work on event tracking springs from the *Topic Detection and Tracking* (TDT) benchmark series, initiated in the late nineties. The aim of TDT was to develop core technologies for news understanding systems [1]. More specifically, its tasks focused on discovering and keeping track of real-world events in multilingual news streams from various sources. The TDT tracking task models the information need of users who hear about a certain event in the news and want to be notified about all follow-up stories. Various methods have been developed for this task, including machine learning and query expansion-based methods and statistical language modeling techniques [7,8].

More recently, the exploding size and popularity of social networks and social media has encouraged researchers to experiment with topic detection and tracking techniques on huge social network data streams such as Twitter [4,11,12]. Because of the real-time and mobile nature of social media, they are particularly suitable for sharing and discussing real-world events or popular tv programs, *as they are happening.* Therefore, one can exploit social media content to identify and keep track of such events, which is exactly the goal of our research.

The speed and volume at which messages are posted on Twitter require techniques which are scalable and robust against imbalanced, sparse and noisy data. [12] addresses the scalability problem of monitoring millions of tweets by using locality-sensitive hashing techniques for detecting the first tweet about an event. [11] builds further on this system, adding Wikipedia as a mechanism to filter out spurious events. [4] proposes an on-line framework which combines content-based clustering with a classifier for separating event from non-event clusters.

TwitterStand is a Twitter-based news understanding system that tries to capture tweets which correspond to late breaking news [14]. It applies content-based clustering and deals with the noisy nature of Twitter by only allowing event clusters that contain tweets from reputable sources. The system automatically discovers the most important hashtags corresponding to an event by simply aggregating and thresholding all hashtags that occur in a cluster. While Twitter-Stand focuses on news events, which are typically short-lived and therefore not subject to significant topical shifts, we are interested in following more prolonged and complex events, like multi-day pop festivals or sports events. Social media activity around such types of events tends to be highly dynamic. We think that capturing the topical evolution of such an event asks for a careful and focused event profile expansion strategy.

In the past few years, query expansion techniques for microblog retrieval have been studied by several researchers. [2] applies query expansion using external resources as a way to cope with the extreme brevity of tweets. The work in [9] uses a time-dependent scoring function for the selection of relevant terms, which are used to expand the original query. The idea behind this is that the language use around a topic continuously evolves. [10] proposes an approach which generates expansion terms based on the temporal co-occurrence of terms. These studies are different from our work in that their objective is to improve search on a microblog *archive* rather than searching for event-related material in a continuous stream of tweets *as they are posted*, and we also look into the concept of interestingness in addition to relevance.

A recent study which is related to ours uses hashtags for query expansion during relevance feedback [5]. Finding the hashtags to follow is basically a more constricted version of our usecase, which is to find keywords, hashtags and usernames that should be followed to track an event. Similar to us, [5] constructs a language model for each hashtag, which is compared to the query language model using Kullback-Leibler (KL) divergence. The 25 top ranked hashtags are then used to create a feedback model based on uniform distributions, or on inverse document frequencies (IDF) of the tags. The feedback model is linearly combined with the original query model. In addition tag co-occurrences are taken into account, to indicate presence in the neighbourhood of the query. Nevertheless, this work is applicable to a static collection of items, whereas we need to deal with continuous streams of tweets. Therefore we further elaborate on [5] and adapt it to our needs of finding relevant and interesting tweets in real time.

3   Event Profiler

Our goal is to design an event profiler that, given a main keyword, determines a list of related keywords (i.e., an event profile) to be monitored during an event. Tweets including these keywords are then evaluated for being displayed along with the live event broadcast content. To get the tweets real-time, we make use of the Twitter Streaming API[3]. This API provides us with a feed of tweets that match one or more of the tracked keywords. The list of tracked keywords is created using the event profiler described below.

The starting point of our event profiler is the main keyword, which can consist of a hashtag, a term or a username, and needs to be created manually. This keyword may be selected as the official hashtag of the event that is intended to be followed or could be obtained from an Electronic Programme Guide or website. The event profiler periodically creates a list of additional keywords to track, where keywords include hashtags, terms and usernames. More specifically, three types of keywords are distinguished:

- The main keyword, which is always followed; tweets containing the main keyword are shown to the user by default.
- Relevant keywords, which are selected as per the procedure outlined below, and followed during a time period t. Tweets containing a relevant keyword are always shown to the user. At the end of the time period t, the relevant keywords are confirmed to be relevant only if they belong to the top k keywords which are most similar to the main one; otherwise, they are discarded.
- Candidate keywords, which are also followed during a time period t, are keywords that may potentially be relevant. Tweets containing a candidate keyword only, and not one of the relevant keywords or the main keyword, are not shown in the application. At the end of the time period t, a candidate keyword is either promoted to a relevant keyword, if it is one the top k most similar to the main keyword, or it is discarded.

To retrieve and select candidate and relevant keywords, the following process is repeated in regular time intervals t:

1. In period t follow the main keyword to retrieve a set of tweets.
2. From the set of tweets containing the main keyword, extract the top n most frequent terms, hashtags and usernames. These terms are the new candidate keywords.
3. In period $t + 1$ follow the main keyword, the relevant keywords and the top n candidate keywords to retrieve a set of tweets.
4. All tweets that contain the main keyword or one of the relevant keywords are considered relevant and are shown to the user
5. For each keyword generate a Maximum Likelihood Estimation language model based on the tweets that contain the keyword [13].

---

[3] https://dev.twitter.com/docs/streaming-apis/streams/public

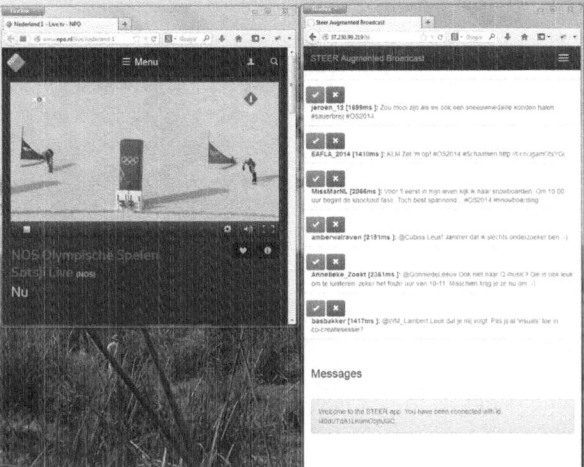

Fig. 1. A screenshot of the social TV application.

6. For each candidate and relevant keyword, assess the similarity of its language model and that of the main keyword. To assess the similarity between two language models we use Kullback-Leibler divergence [17].
7. Select the k keywords (either candidate or relevant) most similar to the main keyword, and designate them as relevant keywords for the upcoming time period.
8. Go to 2.

Note that the system needs two time intervals to start up, in these first two time periods the list of relevant keywords is still empty. The identified relevant tweets are displayed and updated in the social TV application shown in Figure 1.

4   Experiments

In this section we describe our the setup and the outcomes of the experimental validation of our event profiler.

4.1   Experimental Set-up

We tested our event profiler during Sochi Winter Olympics 2014. We performed two sessions at two different locations, each covering live speed skating finals. As our main keyword we used "#os2014" which is the hashtag that was promoted by the television broadcaster. In Figure 1 a screenshot of the social TV application used in our experiment is shown. The application consists of two components: the live television broadcast (lefthand side of the screen) and the real-time tweets and photos stream (on the righthand side). When relevant tweets are retrieved, they are displayed at the top of the tweet stream.

The experiment was set up as follows. After a training session (to let participants get familiar with the application and to understand the task), participants watched 25 minute live broadcast. While watching the broadcast, participants saw tweets and photos appearing on the righthand side of the screen. The participants gave feedback on whether they liked the tweets or not by clicking the green or red button, respectively. All interaction with the application was logged.

As we were interested in checking the added value of using the event profiler to track relevant tweets, and also to check the dependency of its performance on the number of keywords k tracked, we defined three levels of k for this experiment, changing every 5 minutes. In the first setting (k low), no further keywords were selected besides the main one, and $k = 0$. This setting is equivalent to not using the event profiler. In the second setting (k middle), we tracked 4 keywords on top of the main one. Finally, in the last setting (k high), we tracked 10 keywords in total, so $k = 9$.

Since tweets were updated quite frequently during the broadcast, and were also competing for attention with the main event video, users may have failed in evaluating (non-)relevant tweets while the broadcast was displayed. To have a more controlled evaluation of the tweets, uniform across all participants, we asked all our participants to re-evaluate the relevancy and interestingness of a subset of the tweets displayed during the event. Those were randomly selected from the complete set of displayed tweets with the requirement that (1) they would cover all relevant keywords at least once and (2) they had at least one feedback (i.e., like or dislike) during the live broadcast. Tweets were evaluated on a 7-point Likert scale for both interestingness and relevancy. In addition, participants were asked to rate the relevancy of all the relevant keywords generated by the event profiler we used in selecting tweets (again, on a 7-point scale). 41 tweets were rated after the live broadcast in session A and 44 tweets were rated after session B (note that these tweets were necessarily different, since they referred to two different events). In total, participants rated 32 tweets with a low level of relevant keywords, 34 tweets with middle level of relevant keywords and 19 tweets with high level of relevant keywords. Twenty participants (7 female and 13 male, mean age = 28.15) were recruited for the experiment (10 per session).

4.2   Experimental Results

We first analyze the likeability of tweets according to the feedback given by our participants while the event was occurring. We then investigate the relevancy and interestingness of tweets as well as keywords as judged after the event, during our controlled feedback session. Finally, we look at the relationship between interestingness and relevancy of the tweets displayed by our event profiler.

Likes and Dislikes. In total, during the experiments, 5.959 of the clicks evaluated 1.729 different tweets, with an average of 300 clicks per user. The fraction of positive responses per user (likes divided by the sum of likes and dislikes) can be found in Figure 2a. Most users liked between 20% and 40% of the tweets they

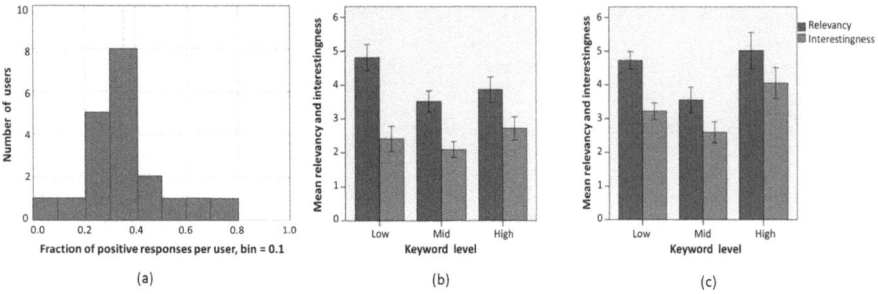

Fig. 2. Results of the user test. (a) positive fraction of liked tweets during the live broadcast; relevancy and interestingness of tweets depending on the keyword level for session A (b) and session B (c); low number of keywords equates to the event profiler to be switched off.

gave feedback on. Currently, all tweets we think are relevant (i.e. contain the main keyword or a relevant keyword) are shown in the application. So, there is a large opportunity for improvement to select tweets users like and find interesting for display in the application.

From the 1.729 tweets that receive a click, 721 contain the main keyword, 1.008 do not.On average the positive fraction for tweets containing the main keyword is 0.42, for tweets without the main keyword the average positive fraction is 0.32. Although the average positive fraction is significantly lower (tested using an unpaired t-test [16]) for tweets retrieved with the automatically added keywords, this is an encouraging result. By adding keywords automatically using our event profiler we can more than double the number of tweets shown in the application without introducing too much noise.

Tweets Relevancy and Interestingness. The tweet likeability analysis pointed out that tweets from larger sets generated by the event profiler at medium and high keyword levels, were liked similar to those generated by tracking the main keyword only. In this section we analyze the results of the controlled feedback session.

To check whether there were any content and/or sample effects, we performed an independent samples Mann-Whitney U test [16] between the relevancy scores collected in the two sessions. Here relevancy is set as dependent variable while sessions are set as independent variable. The mean of relevancy scores of session B is significantly higher than the mean relevancy of Session A ($U = 77578, Z = -3.6, p < 0.001$). A similar test on the interestingness scores shows those of session A to be significantly lower than those of Session B ($U = 72998.5, Z = -4.9, p < 0.001$). These differences maybe caused by either the different content of the broadcast, or different demographics of the user sample. However, more

Table 1. Intraclass correlation between users' relevancy and interestingness judgments

	Relevancy corr.	Interestingness corr.
Session A	0.50	0.38
Session B	0.73	0.50

research is needed to draw conclusions about this discrepancy. To proceed with our analysis, we split the data based on different sessions and test whether the tweets relevancy and interestingness change with different relevant keywords levels.

We ran two k-independent samples Kruskal-Wallis Tests [16] per session. Here the tweets relevancy and interestingness scores are set as dependent variables, while keywords level was set as independent variable. As shown in Figure 2b and Figure 2c, the keyword level significantly influences tweets relevancy ($\chi_2 = 20.419, p < 0.001$ and $\chi_2 = 29.517, p < 0.001$, for session A and B respectively)) and interestingness ($\chi_2 = 9.706, p = 0.008$ and $\chi_2 = 27.266, p < 0.001$, for A and B respectively) in both sessions. The Y-axis represents the mean score of relevancy and interestingness on a 7-point Likert scale while the X-axis represents the three keywords levels. Whereas for relevancy there is no clear trend in scores, we can see the added value of using our event profiler on the interestingness of tweets. Using the profiler with a high number of keywords is clearly beneficial: in both sessions, tweets are scored higher in interestingness than the baseline condition (low number of keywords). This suggests that even better performances may be obtained by setting k even higher, but further research is needed to confirm this hypothesis. Users moderately agree on the relevancy and interestingness of the selected rated tweets. The intraclass correlation [6] between users of the two sessions can be found in Table 1. The correlation on relevancy is higher than the correlation on interestingness. This suggests that interestingness judgments are more susceptible to personal differences than the relevancy ones.

Keywords Relevancy. Users agree well on the relevancy of keywords (with intraclass correlations of 0.84 and 0.81 for the two experimental groups, respectively), which are scored as either very relevant or not relevant at all (see Figure 3). Irrelevant keywords include many usernames of users not directly related to the event. Usernames of participants in the competition or other iceskaters are rated as relevant. Highly relevant terms include hashtags like #sochi2014 and #10km, and keywords like 'schaatsen' (speed skating) and 'kleibeuker' (one of the Dutch female ice skaters). A number of more general terms such as 'zilver' (silver) and 'vrouwen' (women) is judged as relevant by the participants, but these keywords might not make very good keywords to track, since they are very general and may return irrelevant tweets in which these words are used in another context.

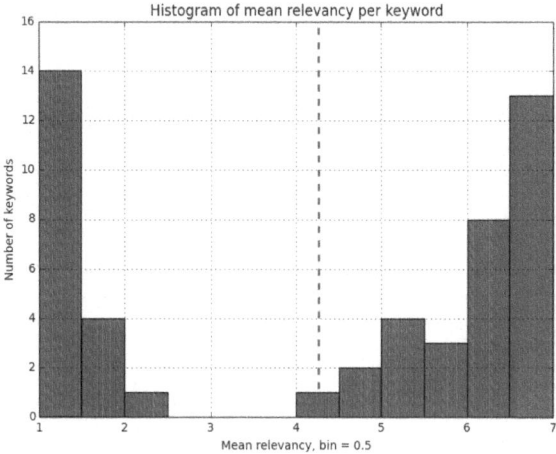

Fig. 3. Histogram of mean relevancy per keyword

Correlation Between Relevancy and Interestingness

To test the relationship between the tweets relevancy and interestingness, we calculated the mean score of relevancy and interestingness of each tweet across all session participants, and computed the Spearman correlation [16] between mean relevancy and interestingness among all tweets (Session A: $r = 0.698, p < 0.001$; Session B: $r = 0.753, p < 0.001$). For individual tweets, interestingness is significantly correlated with relevancy, and the relevancy of tweets is rated consistently higher than the interestingness. To further investigate this relationship, a regression analysis [16] is considered per session. Here the interestingness score is set as dependent variable while relevancy is set as independent variable. A linear, logarithmic and inverse model are used for curve estimation. As shown in Figures 4a and Figure 4b, interestingness is significantly correlated with relevancy in both sessions. The inverse model has the best performance (Session A: $R^2 = 0.526, F(1, 43) = 47.624, p < 0.001$, Session B: $R^2 = 0.638, F(1, 43) = 75.821, p < 0.001$), summarizing indeed the observed relationship: relevant tweets are not necessarily interesting, but interesting tweets are always relevant. The least interesting tweets include tweets like 'Finished learning, watching iceskating now', which are of limited interest for the general public who does not know the author of the tweet. The most interesting tweets include jokes, and facts like the racing times of the Dutch participants, and the announcement of coming races. There are only a few tweets that are judged as more relevant than interesting, these are the points in the lower right corner in Figures 4a and 4b. These tweets are about other events at the Olympic games such as ice hockey, or about the Olympic games in general.

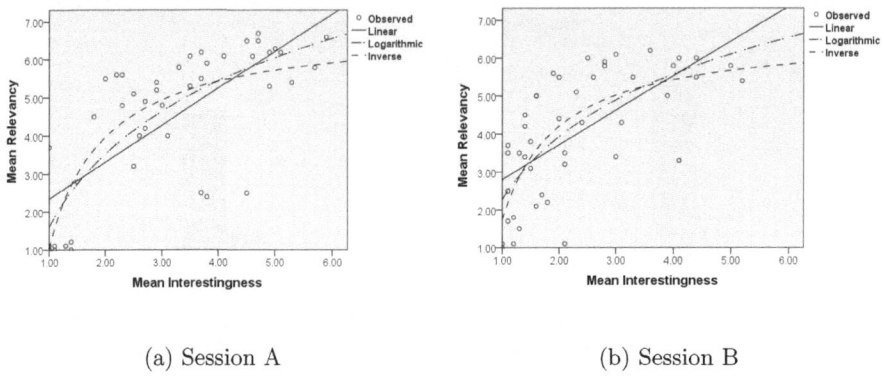

(a) Session A   (b) Session B

Fig. 4. Scatterplot of relation between interestingness and relevancy of tweets.

5   Conclusion

In this paper we have presented an event profiler that retrieves in real time a high number of interesting tweets related to a live event broadcast. The event profiler ranks keywords based on the similarity of their language models to the language model of a manually selected main keyword. From our experimental results we conclude that our event profiler is capable of selecting keywords which lead to retrieval of significantly more likeable tweets than following a single keyword without introducing too much noise. Relevancy and interestingness are found to be correlated, although an inverse function is found to model the data more properly: relevant tweets are not necessarily interesting, but interesting tweets are usually relevant.

In future work we would like to investigate whether we can predict the interestingness of tweets in general, and whether we can optimize the event profiler based on user preferences and user contacts in social networks to increase the interestingness of the tweets shown in the application.

Acknowledgments. Part of the research leading to these results has received funding from the European Union's Seventh Framework Programme (FP7/2007-2013) under grant agreement no. 318343.

References

1. Allan, J.: Topic detection and tracking: event-based information organization, vol. 12. Kluwer Academic Publishers (2002)
2. Bandyopadhyay, A., Ghosh, K., Majumder, P., Mitra, M.: Query expansion for microblog retrieval. International Journal of Web Science **1**(4), 368–380 (2012)
3. Basapur, S., Mandalia, H.M., Chaysinh, S., Seok Lee, Y., Venkitaraman, N., Metcalf, C.J.: FANFEEDS: evaluation of socially generated information feed on second screen as a TV show companion. In: EuroITV 2012, pp. 87–96 (2012)

4. Becker, H., Naaman, M., Gravano, L.: Beyond trending topics: Real-world event identification on twitter. In: Proceedings of the Fifth International AAAI Conference on Weblogs and Social Media (ICWSM 2011) (2011)
5. Efron, M.: Hashtag retrieval in a microblogging environment. In: Proceedings of the 33rd International ACM SIGIR Conference on Research and Development in Information Retrieval, pp. 787–788. ACM (2010)
6. Fisher, R.A.: Statistical methods for research workers. Genesis Publishing Pvt. Ltd. (1925)
7. Kraaij, W., Spitters, M.: Language models for topic tracking. In: Croft, B. (ed.)Language Modeling for Information Retrieval, pp. 95–124. Springer (2003)
8. Lavrenko, V., Allan, J., DeGuzman, E., LaFlamme, D., Pollard, V., Thomas, S.: Relevance models for topic detection and tracking. In: Proceedings of the Second International Conference on Human Language Technology Research, pp. 115–121. Morgan Kaufmann Publishers Inc. (2002)
9. Massoudi, K., Tsagkias, M., de Rijke, M., Weerkamp, W.: Incorporating query expansion and quality indicators in searching microblog posts. In: Clough, P., Foley, C., Gurrin, C., Jones, G.J.F., Kraaij, W., Lee, H., Mudoch, V. (eds.) ECIR 2011. LNCS, vol. 6611, pp. 362–367. Springer, Heidelberg (2011)
10. Metzler, D., Cai, C., Hovy, E.: Structured event retrieval over microblog archives. In: Proceedings of the 2012 Conference of the North American Chapter of the Association for Computational Linguistics: Human Language Technologies, pp. 646–655. Association for Computational Linguistics (2012)
11. Osborne, M., Petrovic, S., McCreadie, R., Macdonald, C., Ounis, I.: Bieber no more: First story detection using twitter and wikipedia. In: SIGIR 2012 Workshop on Time-Aware Information Access (2012)
12. Petrović, S., Osborne, M., Lavrenko, V.: Streaming first story detection with application to twitter. In: Human Language Technologies: The 2010 Annual Conference of the North American Chapter of the Association for Computational Linguistics, pp. 181–189. Association for Computational Linguistics (2010)
13. Ponte, J.M., Croft, W. B.: A language modeling approach to information retrieval. In: Proceedings of the 21st Annual International ACM SIGIR Conference on Research and Development in Information Retrieval, pp. 275–281. ACM (1998)
14. Sankaranarayanan, J., Samet, H., Teitler, B.E., Lieberman, M.D., Sperling, J.: Twitterstand: news in tweets. In: Proceedings of the 17th ACM SIGSPATIAL International Conference on Advances in Geographic Information Systems, pp. 42–51. ACM (2009)
15. Schirra, S., Sun, H., Bentley, F.: Together alone: motivations for live-tweeting a television series. In: Proceedings of the 32nd Annual ACM Conference on Human Factors in Computing Systems, pp. 2441–2450. ACM (2014)
16. Sheskin, D.J.: Handbook of parametric and nonparametric statistical procedures. CRC Press (2003)
17. Zhai, C., Lafferty, J.: Model-based feedback in the language modeling approach to information retrieval. In Proceedings of the tenth International Conference on Information and Knowledge Management, pp. 403–410. ACM (2001)

Topic Detection in Twitter Using Topology Data Analysis

Pablo Torres-Tramón$^{(\boxtimes)}$, Hugo Hromic, and Bahareh Rahmanzadeh Heravi

Insight Centre for Data Analytics, National University of Ireland, Galway, Ireland
{pablo.torres,hugo.hromic,bahareh.heravi}@insight-centre.org

Abstract. The massive volume of content generated by social media greatly exceeds human capacity to manually process this data in order to identify topics of interest. As a solution, various automated topic detection approaches have been proposed, most of which are based on document clustering and burst detection. These approaches normally represent textual features in standard n-dimensional Euclidean metric spaces. However, in these cases, directly filtering noisy documents is challenging for topic detection. Instead we propose Topol, a topic detection method based on Topology Data Analysis (TDA) that transforms the Euclidean feature space into a *topological space* where the shapes of noisy irrelevant documents are much easier to distinguish from topically-relevant documents. This topological space is organised in a network according to the connectivity of the points, i.e. the documents, and by only filtering based on the size of the connected components we obtain competitive results compared to other state of the art topic detection methods.

1   Introduction

Social Network Sites (SNS) are one of the most important communication channels nowadays. SNS users interact with one another generating a considerable amount of content of various media types such as text, images or videos. This content has the potential of reaching a very wide audience, where feelings, political opinions or breaking news can be transmitted. One particular kind of SNS are *microblogging* sites, where messages are constrained and normally rather short. Twitter is the prime world-wide example of a microblogging system. In this environment, information is shared and circulated faster than in more conventional SNS such as blogs or forums, reaching a large audience in a shorter time.

Real-world events have shown the key role of microblogs for spreading news and supporting the information flow between communities in the social sphere. For example, the Mumbai 2008 bomb blasts, the 2011 crash of the US Airways Flight 1549, the Arab Spring movements, and the Boston Marathon bombing were all very important global events where social media played a crucial role in reporting and covering the news [9]. In such situations, users acted as *real-life sensors* [5], reporting what was happening nearby and posting information almost in real-time. All of this content can be mined in order to explore and monitor real-world events. In particular, we are interested on detecting related

© Springer International Publishing Switzerland 2015
F. Daniel and O. Diaz (Eds.): ICWE 2015 Workshops, LNCS 9396, pp. 186–197, 2015.
DOI: 10.1007/978-3-319-24800-4_16

topics inside the context of a larger story. We want to identify related stories that may not have been previously considered, and hence enrich the main story itself. This use case is key within a journalism context where the journalist is concerned about all the details for a particular story [1].

Numerous research studies have been conducted to create methods that automatically detect topics in real-world events such as government crises [15], natural disasters [18] or political elections [6]. Most of these methods use ranking or clustering to determine whether a topic is of relevance or not. Clustering, for instance, requires defining a linkage strategy and a series of thresholds to select candidates. A similar situation occurs for ranking-based methods because they need to select a subset of the highest candidates. Even if clustering and ranking approaches are suitable and have good results for a large number of use cases, they often require to repeatedly train the model when facing new data in order to calibrate the thresholds. This requirement makes topic detection methods too rigid for the context of breaking news analysis in microblogs systems.

In this paper we propose TOPOL, a novel unsupervised method for detecting topics in Twitter data based in Topology Data Analysis (TDA). The fundamental goal of TDA is to recognise shapes or patterns present within the data [14]. TDA defines a coordinate system, the *topological space*, generated using a *distance* function and transforms the input data so that this new space does not consider coordinates but distances instead. The central idea of topology analysis is the fact that it allows studying the properties of data shapes that are invariable under small deformations [14]. In addition, TDA also allows to study different perspectives of the same data. Our solution *explores the shapes* formed by Twitter data represented as a network of overlapping clusters, and uses this graph to determine underlying topics. Our intuition is that major topics are concentrated on large and densely connected components within this network. On the other hand, noisy topics are represented as small and isolated groups of nodes.

In our experiments, TOPOL shows to be competitive compared to state of the art topic detection methods for the same use case. While current approaches rely mostly on clustering and filtering techniques, our method identifies topics and generates their descriptions only from the shapes of the data alone, according to the constructed topological spaces using TDA.

The remainder of the paper is organised as follows. In Section 2 we provide a brief description of the Twitter topic finding problem that we address in this work. Section 3 discusses current state of the art methods for the above task. In Section 4 we describe the general TDA approach and in Section 5 we introduce our algorithm, TOPOL. Section 6 describes the experiments and results, and Section 7 concludes the paper and provides future interesting directions for our research. Because we focus on the Twitter context, from now on we will use the terms *tweets*, *post*, and *documents* interchangeably.

2   Problem Description

We address the problem of topic detection in Twitter. This task can be defined as identifying prominent topics in a document corpusunder a User-centred scenario [2], where the documents in this case are *tweets* posted in Twitter. Since Twitter data is continuously generated, the aforementioned document collection is then inherently stream-based and suitable approaches require constantly updating their output according to newly arriving data items in order to incorporate the latest changes, i.e. new tweets being created.

One mechanism to handle streams of documents is the usage of sliding windows techniques. This scheme defines an *update rate*, which in turn creates *time slots* as the time period between each update. The value for this update rate parameter is dependent on the nature of the event under consideration. For example, if the event continues for a few minutes the time slots should be small, but in contrast, if the event lasts for days, the time slots period should be larger. We then refine the topic finding problem to identifying topics in each of those time slots or windows. Furthermore, we represent the discovered topics as a list of keywords and a satisfactory detection of topics should bring the most representative keywords for each of them.

The content of tweets normally includes a wide range of subjects, such as personal feelings, political opinions, breaking news information, spam or comments. Such variety imposes difficult challenges and complexities for the topic detection task. In order to frame the experiments described in this paper, the input data is narrowed down by predefining a set of keywords such that every tweet must content at least one of those keywords. This a priori information is considered to be provided by the end-user and the keywords are assumed to be highly related to some event of interest for studying.

3   Related Work

Topic detection on large streaming data, such as Twitter, gained notorious interest by researchers in the last few years. There are two main branches of approaches: (1) *document-pivot* where the topics are identified from the documents and (2) *feature-pivot* where the topics instead are generated according to relations found in a diverse range of features.

An example of a document-pivot approach can be found in [16]. The authors address topic detection in Twitter by clustering documents (tweets). Because generating clusters is a time-consuming task, they implemented a more efficient method by using Local Sensitive Hashing (LSH). This improvement allows to find the nearest clusters for a new document in constant time, dramatically reducing the computational effort for document comparison. Additionally, in order to reduce non-relevant topics, they established *threads* of topics such that each thread corresponds to the evolution of a particular topic across time. This information is used to filter out non-interesting topics. However, this method still is a form of clustering and hence it suffers from data fragmentation. In contrast,

TOPOL groups documents together according to the connections present in the feature spaces, found by repeatedly sampling the tweets being analysed.

Feature-pivot methods rely on finding associations in a subset of defined features. The goal of these approaches is to (1) reduce the computation time by considering only a subset of features and (2) improve the topic detection results by only using a higher score for this subset of features. Several strategies have been proposed to *identify a suitable subset of features* such as probabilistic models [7], ranking [20] or Wavelet analysis [22]. Once features are selected, they are *analysed* in order to extract associations to later build topics. For this there are several techniques as well, such as clustering [8], ranking [1] or noise reduction [10].

Selecting feature subsets contribute to reducing non-relevant topics, but also create a bias in the final output depending on the selection criteria. Our algorithm, TOPOL, does not require selecting features but instead studies the topological shapes of the data directly according to a chosen similarity function.

It is also common to find a combination of strategies [1,10]. In general, feature-pivot approaches tend to generate misleading correlations between features and found topics that in reality are not associated with any event of interest [3,11].

Finally, Sayyadi et al. [20] proposed a similar approach to TOPOL. The authors represented the document features in a graph of keywords where nodes are terms and the links between them are the co-occurrence degrees of two terms in the tweets. Afterwards, topics are determined by community structures within this graph. This method differs from our approach since TOPOL builds a network according to a certain distance function instead. This particularity makes it possible for one feature to co-occur in several documents but, if they are not *close* in a topological sense, they will be not associated with the same topic.

4 Topology Data Analysis (TDA)

In many topic detection approaches, a text stream is traditionally represented using a vector where each feature corresponds to one coordinate in a Cartesian system. The similarity (or dissimilarity) of those vectors is defined using a distance function such as the well-known Euclidean-distance or Cosine-similarity functions. Moreover, this distance function is assumed to be continuous for all text streams, which means that it is always possible to define a distance between any two documents. However, these assumptions are far from being realistic in most real-world use cases.

For the above reason, instead of assuming a Cartesian system, it is preferable to study the data without considering the raw underlying metric space, therefore reducing the background noise embedded within this coordinate system. For this purpose, we use Topology Data Analysis (TDA) to generate representations of the data that allow us to study the inherent invariant shapes within this data. TDA is rooted on the field of *Topology*, which is the branch of Mathematics that deals with *qualitative* instead of quantitative information [4]. In addition,

Topology is coordinate-free, which means it studies the geometric properties of the data without depending on any particular coordinate system, and uses the notion of *infinite nearness* instead of a distance function.

In this work, we employed the MAPPER algorithm [21] to generate topological representations of Twitter data. This method is based on a generalised version of Reeb graphs [17]. MAPPER, as suggested by its name, applies a *mapping function* to construct a network-based representation of the input data points. This input is first valued according to a *distance function*. The algorithm iteratively samples the constructed distance matrix in small subsets of points that are evaluated by a *filter function*, whose image is further divided into intervals that are related to those subsets.

The aforementioned distance function is the core mathematical tool that characterises each point in the feature space. The interval size parameter, called the *resolution* and denoted as r_p, is variable. With bigger intervals a more general vision of the data can be obtained. On the other hand, if the intervals shrink, the generated output is built according to the smaller shapes of the input data. It can be noted that this size parameter determines the amount of intervals used by the filter function. The points assigned to the same interval can be considered as partial *clusters*, which later corresponds to a node in the output network.

The graph generated is a representation of the connection of the points in the space, the mapping function is designed to intentionally overlap the intervals to some degree, allowing for a bunch of points to co-occur in between a group of intervals. This number of occurrences among the intervals reflects how connected the points are in the space. An *overlapping* parameter, denoted as o_p, is then defined that ranges between 0% and 100%. This value controls the overlapping degree, with a larger value meaning that there will be a greater probability for the same points to lie in two or more intervals.

The final output of the MAPPER algorithm is a network-based representation of the input data such that each partial cluster is a node and if two partial clusters have one or more shared points – according to the overlapping intervals – the nodes are linked together. Figure 1 shows a toy example of this output using a 3-dimensional synthetic input dataset. This dataset is a collection of points that resemble two touching spheres (Figure 1(a)). After generating the graph representation of the partial clusters using MAPPER, we obtain the network shown in Figure 1(c), assuming an Euclidean distance. The colours in this network represent the output values of the filter function associated with each partial cluster (node).

If we now add extra noise points to the synthetic dataset (Figure 1(b)) the generated network now represents those noisy points in isolated nodes – as shown in Figure 1(d) – because they can be easily separated as such from a topological perspective. Moreover, these two independent datasets in this space are represented as clearly isolated components in the network thus enabling them to be studied separately.

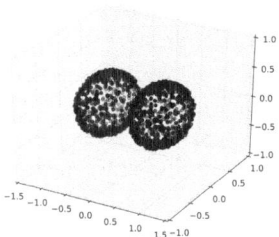

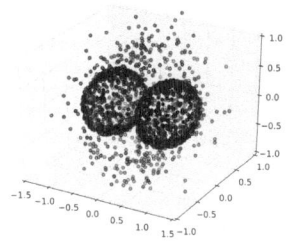

(a) Clean three-dimensional input dataset containing two touching spheres

(b) Noisy three-dimensional dataset with randomly added noise points

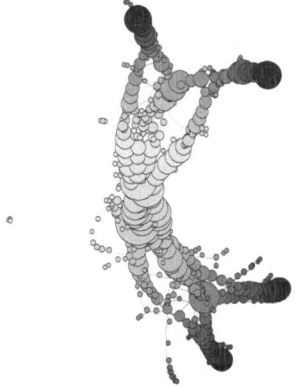

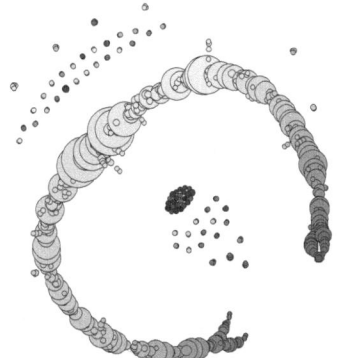

(c) Output network from the MAPPER algorithm for the clean dataset

(d) Output network from the MAPPER algorithm for the noisy dataset

Fig. 1. Example network outputs for the *mapper* function. A clean input dataset 1a is represented as a graph that describes how the points are connected 1c. For the case of a noisy dataset 1b, the output graph models the noise as isolated nodes 1d. The colours in the networks show the output values of the *filter* function for each node.

5   Topol: A TDA-Based Topic Detection Approach

We now provide an overview of TOPOL, our topic detection algorithm for Twitter based on Topology Data Analysis. We divide our method in three steps: *pre-processing*, *mapping* and *topic detection*. All of those are described below.

5.1   Pre-Processing Step

In TOPOL, we represent documents (the tweets) as a *bag of words* weighted by the standard Terms frequency (TF) and Inverse Document Frequency (IDF) measures [19]. Furthermore, each document has a timestamp associated indicating the moment when it was created.

We then use the windowing scheme described in Section 2 and for each time slot we perform a cleansing and filtering processes since the data still can contain undesired posts such as spam. For this we follow a similar strategy as suggested by Ifrim et al. [10]. This approach assumes that a tweet is noisy if the number of *user mentions* or *hashtags* (user-provided tags) are above a defined threshold. Even though this strategy is very simple, Ifrim et al. reported that it effectively reduces noisy tweets, specially spam and advertising since posts in these categories tend to have a high number of user mentions and hashtags. For our experiments we decided to set a conservative filtering threshold of 2, leading to 20% of the input tweets being removed.

For extracting the features we will later study using TDA from each tweet, we employ the following approach: first, we eliminate all the *URLs, hashtags, user mentions* and any non-textual symbols for all the remaining tweets. Later, all non-ASCII characters are further removed as well as punctuation marks, digits and stop words. The remaining text for each tweet is then tokenised according to white spaces and a *TF-based vector* is generated to represent the tweet. Finally we perform an additional filtering by only selecting those TF vectors that have at least more than four distinct terms or features. In summary, at the end of this pre-processing step, each selected tweet is represented as a TF vector using a globally kept dictionary.

5.2   Mapping Step

After pre-processing, we apply the MAPPER algorithm to the TF vectors in the current time slot. for this, first it is necessary to generate an input metric space. Therefore, we compute the *all-to-all* distance matrix M for all the tweets in the window. For the required filter function, we generate a rectangular diagonal matrix by applying the standard Singular Value Decomposition (SVD) technique to the distance matrix M. The values of this function are then used for sampling the distance matrix. The resolution (r_p) and overlapping (o_p) parameters are set to different values in our experiments to obtain a variety of network-based representations of the TF vectors that model the input tweets (see Section 6).

MAPPER divides the input space using the following work-flow: (1) it selects the maximum and minimum values of the filter function, (2) it calculates the length of the intervals according to the resolution parameter r_p, and (3) the intervals I_i are set such that they overlap using the overlapping parameter o_p. For example, if $o_p = 50\%$ the resulting intervals will share half of the available space as follows:

$$I_0 = [x_0, x_1]$$
$$I_1 = [x_1 - r_p * 0.5, x_1 + r_p * 0.5]$$

Note that all possible intervals in the image of the filter function are covered, starting from the minimum value found to the maximum. In other words, for each interval I_i, MAPPER selects points such that the image of those points lie in the interval I_i. When there are enough points (> 5) in an interval, the algorithm

performs clustering using Single-linkage Clustering [12]. After this, each cluster
is modelled as a node in the output network of MAPPER. If one or more of the
selected points are already assigned to a different node (i.e. cluster), MAPPER
creates a link between them in the output network.

5.3   Topic Detection Step

To this point, the network-based representation generated with MAPPER for each
window represents the data in the feature space according to the filter function
for that particular time slot. Since the noisy tweets tend to create isolated nodes
in this network, the most *relevant* connected components are good candidates
for identifying interesting topics. Furthermore, their most common features can
be used as the topic descriptions.

   Therefore, we define the topics we are interested in as the connected components in the resulting network such that the number of tweets associated with
the component is above a defined threshold α. On the other hand, we use the
β-most frequent features in the same components as their descriptions.

   Our proposed process for identifying topics is performed independently on
each time slot. Once all time slots are processed, we track similar topics across
all the time windows by measuring the Cosine similarity between the topics in
the current and preceding time slots. For this we create independent time series
for each topic such that the topic does not match any other topic according to
the similarity function. For example, if the topic t_0 is present in the windows
w_0, w_1, w_2 and the topic t_1 only in the window w_1, we generate two independent
time series as follows:

$$ts_0 = \{tf(t_0), tf(t_1), tf(t_2)\}$$
$$ts_1 = \{0, tf(t_1), 0\}$$

6   Experiments and Results

To evaluate TOPOL we use the same evaluation framework proposed by Aiello
et al. in [1], where they studied three major real-world events that occurred in
2012. We selected one in particular, the *FA Cup Final*, to conduct our experiments. The FA Cup Final is the final match of the Football Association Challenge
Cup played by the Chelsea FC and Liverpool FC teams on May 5th of 2012.
Chelsea won the match with a final score of 2-1. A set of keywords and *hashtags* provided by experts was used to retrieve related posts from Twitter. The
identifiers of those tweets are all publicly available[1].

   We retrieved the tweets using the Twitter REST API[2]. The dataset was
partitioned in time slots considering the nature of the event (using time slots
corresponding to 1 minute). Aiello generated a ground truth for the dataset
consisting of a manual review of published media reports about the event. This

---

[1] http://www.socialsensor.eu/results/datasets/72-twitter-tdt-dataset
[2] https://dev.twitter.com/rest/public

gold-standard includes 13 topics: the goals scored by players Ramirez, Drogba and Carrol respectively, as well as the kick-off, half-time and the end of the match, among others. According to Aiello, the stories selected were *"significant, time-specific and well represented on news media"*. The start time assigned for each story corresponds to the time that the story emerged in mainstream news. To compare our own results we use the same metrics proposed by Aeillo et al. in their work:

Topic Recall (T-REC) is the percentage of ground truth topics correctly detected by the method. A topic is successfully detected if the keywords that comprise the topic description and the keywords mentioned in the ground truth description have a Levenshtein similarity $>= 0.8$ (as defined by Aiello).

Keyword Precision (K-PREC) is the percentage of successfully detected keywords in the topic description over the total keywords found by the method for the topic description.

Keyword Recall (K-REC) is the percentage of successfully detected keywords for the topic description over the total keywords included in the topic description of the ground truth.

Since there are many other topics in the dataset that are not described in the ground truth, it is not possible to calculate the true *Topic Precision*. More information about this dataset can be found in [1].

Table 1 shows the maximum T-REC and K-REC values achieved for different configurations of TOPOL. We evaluated a wide range of values for the tunable parameters, including the distance function, resolution and overlap as well as

Table 1. Comparison of Topic Recall (T-REC) and Keyword Recall (K-REC) for different distance functions, resolutions (r_p) and overlapping degrees (o_p).

T-REC for Euclidean distance				T-REC for Cosine similarity			
Res (r_p)	Overlapping (o_p)			Res (r_p)	Overlapping (o_p)		
	25	50	75		25	50	75
5	0.385	0.462	**0.538**	5	0.231	0.385	0.385
10	0.308	0.308	0.462	10	0.308	0.308	**0.462**
25	0.231	0.154	0.308	25	0.231	0.308	0.308
50	0.231	0.231	0.231	50	0.308	0.308	0.308

K-REC for Euclidean distance				K-REC for Cosine similarity			
Res (r_p)	Overlapping (o_p)			Res (r_p)	Overlapping (o_p)		
	25	50	75		25	50	75
5	0.571	0.667	0.643	5	0.571	0.600	0.600
10	**0.714**	0.529	0.583	10	0.556	0.526	0.591
25	0.500	0.500	0.692	25	0.556	0.600	**0.667**
50	0.600	0.571	0.600	50	0.600	0.600	0.600

other parameters. Surprisingly, the Euclidean distance function has the better T-REC on average than the Cosine similarity, as opposed to the intuition that Cosine similarity is better suited for text documents. However, since the length of the tweets in Twitter is relatively short and pretty much constant, the Euclidean distance can distinguish elements better than the Cosine similarity. This explains why the performance of our method increases when using the Euclidean distance.

We also observe that Topic Recall increases when the overlapping degree grows, suggesting that MAPPER requires an increased sampling in order to generate better connected components in the output network. This in turn suggests that the tweets are fairly scattered in the space independently of the distance function used for the mapping process. Therefore the connected components cannot be easily linked together in the network if we use a low overlapping value.

In contrast, when the resolution increases the Topic Recall metric decreases. With higher resolutions, the generated networks will have more nodes since the intervals of the filter function will be shorter. This creates networks with few connected components and this reflects the high separability of Twitter data at smaller levels, preventing a too connected network. Since we assume that small connected components in the output network are correlated to noise, in this scale the number of candidate topics becomes nearly zero. This observation explains the low Topic Recall obtained.

We studied the influence of the α and β parameters for selecting and describing topics by modifying their values while keeping the other parameters constant (see Figure 2). In this experiment, Topic Recall remained almost unchanged. This indicates that TOPOL benefits greatly from the Topology Data Analysis (TDA) mapping process, and even more than from the burst-based topic descriptions event detection approach.

Finally, we compared TOPOL with state-of-the-art methods studied by Aiello et al. in [1]. We found that our method has competitive results as seen in Table 2.

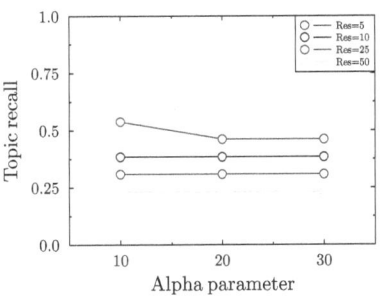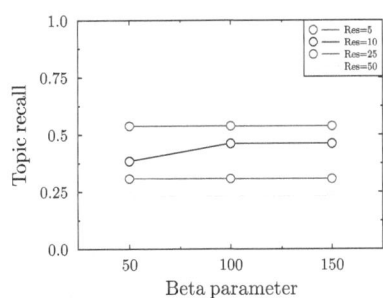

Fig. 2. Topic Recall (T-REC) for different values of α (with a fixed $\beta = 50$), β (with a fixed $\alpha = 10$) and sampling resolution (Res) parameters. The remaining parameters are maintained invariable.

Table 2. Comparison of state-of-the-art topic detection methods studied by Aiello et al. [1] and TOPOL using the Euclidean distance, $r_p = 5$ and $o_p = 75$ as parameters.

Topic Detection Method	T-REC	K-PREC	K-REC
Latent Dirichlet Allocation (LDA)	0.6923	0.1637	**0.6829**
Document-pivot	**0.7692**	0.3373	0.5833
Frequent Pattern Mining (FPM)	0.3077	**0.7500**	0.4286
Soft Frequent Pattern Mining (SFPM)	0.6154	0.2336	0.6579
BNgram	**0.7692**	0.2989	0.5778
TOPOL (based on TDA)	0.5380	0.3000	0.6430

7   Conclusions and Future Work

Detecting events in Social Network Sites (SNS) is a complex process that demands a combination of techniques such as data mining, information retrieval and text mining in order to find stories of interest that are trending in the SNS.

We introduced TOPOL, a novel method to detect topics in Twitter using Topology Data Analysis (TDA). Our method generates a network-based representation of Twitter posts that correlates with the topological shape of keywords modelled as term frequency (TF) vectors according to different distance functions. We evaluated our approach with a standard dataset and distance methods, obtaining competitive results compared to using state-of-the-art approaches [1].

We also found that the most influential parameters for our method are the overlapping degree (o_p) and the sampling resolution (r_p). Both parameters provided significant improvements in our evaluation metrics, specially Topic Recall (T-REC). In addition we showed that TOPOL relies mostly on the usage of TDA than the selection of features, improving the robustness of our approach.

Several future directions can be considered in order to improve the performance and quality of our topic detection method. Many other alternatives to the MAPPER algorithm have been developed in recent years [13]. These new outcomes avoid filtering functions and improve the computational performance. Additionally, the study of the effects of other distance functions is promising. Furthermore, different approaches can be explored as well for detecting topic changes in the topological networks, and also other algorithms for detecting bursty topics in the time series. Finally, more representational models for the SNS documents can be considered for potentially improving our initial results.

References

1. Aiello, L.M., et al.: Sensing trending topics in Twitter. IEEE Transactions on Multimedia **15**(6), 1268–1282 (2013)
2. Allan, J.: Topic Detection and Tracking: Event-based Information Organization, vol. 12. Springer Science & Business Media (2002)
3. Atefeh, F., et al.: A Survey of Techniques for Event Detection in Twitter. Computational Intelligence (2013)

4. Carlsson, G.: Topology and Data. Bulletin of the American Mathematical Society **46**(2), 255–308 (2009)

5. Castillo, C., et al.: Information credibility on twitter. In: Proc. of WWW, pp. 675–684. ACM (2011)

6. Conover, M., et al.: Political polarization on twitter. In: Proc. of ICWSM, AAAI (2011)

7. Fung, G.P.C., et al.: Parameter free bursty events detection in text streams. In: Proc. of VLDB, pp. 181–192. VLDB Endowment (2005)

8. He, Q., et al.: Bursty feature representation for clustering text streams. In: Proc. of SDM, pp. 491–496. SIAM (2007)

9. Heravi, B.R., et al.: Introducing Social Semantic Journalism. The Journal of Media Innovations **2**(1), 131–140 (2015)

10. Ifrim, G., et al.: Event Detection in Twitter using Aggressive Filtering and Hierarchical Tweet Clustering. In: SNOW-DC @ WWW, pp. 33–40. ACM (2014)

11. Imran, M., et al.: Processing Social Media Messages in Mass Emergency: A Survey. arXiv preprint arXiv:1407.7071 (2014)

12. Jain, A.K., et al.: Algorithms for Clustering Data, vol. 6. Prentice Hall, Englewood Cliffs (1988)

13. Liu, X., et al.: A Fast Algorithm for Constructing Topological Structure in Large Data. Homology, Homotopy and Applications **14**(1), 221–238 (2012)

14. Lum, P., et al.: Extracting insights from the shape of complex data using topology. Scientific Reports 3 (2013)

15. Panisson, A.: Visualization of Egyptian revolution on Twitter (February 2011). https://www.youtube.com/watch?v=2guKJfvq4uI

16. Petrović, S., et al.: Streaming first story detection with application to Twitter. In: Proc. of HLT, pp. 181–189. ACL (2010)

17. Reeb, G.: Sur les points singuliers d'une forme de Pfaff completement intégrable ou d'une fonction numérique. CR Acad. Sci. Paris **222**, 847–849 (1946)

18. Sakaki, T., et al.: Earthquake shakes Twitter users: real-time event detection by social sensors. In: Proc. of WWW, pp. 851–860. ACM (2010)

19. Salton, G., et al.: Term-weighting Approaches in Automatic Text Retrieval. Information Processing & Management **24**(5), 513–523 (1988)

20. Sayyadi, H., et al.: Event detection and tracking in social streams. In: Proc. of ICWSM. AAAI (2009)

21. Singh, G., et al.: Topological methods for the analysis of high dimensional data sets and 3D object recognition. In: Proc. of SPBG, pp. 91–100. IEEE (2007)

22. Weng, J., et al.: Event detection in twitter. In: Proc. of ICWSM, pp. 401–408. AAAI (2011)

Author Index